Lecture Notes in Computer Science 10140

Commenced Publication in 1973
Founding and Former Series Editors:
Gerhard Goos, Juris Hartmanis, and Jan van Leeuwen

Editorial Board

More information about this series at http://www.springer.com/series/8637

Abdelkader Hameurlain · Josef Küng
Roland Wagner · Tran Khanh Dang
Nam Thoai (Eds.)

Transactions on Large-Scale Data- and Knowledge- Centered Systems XXXI

Special Issue on Data and Security Engineering

 Springer

Editors-in-Chief

Abdelkader Hameurlain
IRIT, Paul Sabatier University
Toulouse
France

Roland Wagner
FAW, University of Linz
Linz
Austria

Josef Küng
FAW, University of Linz
Linz
Austria

Guest Editors

Tran Khanh Dang
Ho Chi Minh City University of Technology
Ho Chi Minh City
Vietnam

Nam Thoai
Ho Chi Minh City University of Technology
Ho Chi Minh City
Vietnam

ISSN 0302-9743   ISSN 1611-3349 (electronic)
Lecture Notes in Computer Science
ISSN 1869-1994   ISSN 2510-4942 (electronic)
Transactions on Large-Scale Data- and Knowledge-Centered Systems
ISBN 978-3-662-54172-2   ISBN 978-3-662-54173-9 (eBook)
DOI 10.1007/978-3-662-54173-9

Library of Congress Control Number: 2016961429

Printed on acid-free paper

This Springer imprint is published by Springer Nature
The registered company is Springer-Verlag GmbH Germany
The registered company address is: Heidelberger Platz 3, 14197 Berlin, Germany

Preface

The Second International Conference on Future Data and Security Engineering (FDSE) and the 9th International Conference on Advanced Computing and Applications (ACOMP) were held in Ho Chi Minh City, Vietnam, during November 23–25, 2015. FDSE is an annual international premier forum designed for researchers and practitioners interested in state-of-the-art and state-of-the-practice activities in data, information, knowledge, and security engineering to explore cutting-edge ideas, to present and exchange their research results and advanced data-intensive applications, as well as to discuss emerging issues on data, information, knowledge, and security engineering. ACOMP annual events focus on advanced topics in computer science and engineering. More specifically, ACOMP solicits papers on security, information systems, software engineering, enterprise engineering, embedded systems, high-performance computing, image processing, visualization, and other related fields.

We invited the submission of both original research contributions and industry papers. For FDSE 2015, we received 88 submissions and, after a careful review process, only 23 papers (20 full and three short ones) were selected for presentation. The call for papers of ACOMP 2015 resulted in the submission of 62 papers, of which we selected only 24 papers through a rigorous review process for presentation.

Among the papers for both FDSE and ACOMP 2015, we selected only seven papers to invite the authors to revise, extend, and resubmit their work for publication in this special issue. Finally, only six extended papers were accepted. The main focus of this special issue is on advanced computing matters in data and security engineering as well as their promising applications.

The huge success of FDSE and ACOMP 2015 as well as this special issue of *Transactions on Large-Scale Data and Knowledge-Centered Systems* (TLDKS) was the result of the efforts of many people, to whom we would like to express our gratitude. First, we would like to thank all authors who extended and submitted papers to this special issue. We would also like to thank the members of the committees and external reviewers for their timely reviewing and lively participation in the subsequent discussion in order to select such high-quality papers to be published in this issue. Last but not least, we thank Gabriela Wagner for her enthusiastic help and support during the preparation of this publication.

October 2016

Tran Khanh Dang
Nam Thoai

Organization

Editorial Board

Reviewers

Lam-Son Lê	HCMC University of Technology, VNUHCM, Vietnam
Nhien-An Le-Khac	University College Dublin, Ireland
Quoc Viet Hung Nguyen	University of Queensland, Australia
Van Doan Nguyen	Japan Advanced Institute of Science and Technology, Japan
Viet Hung Nguyen	Trento University, Italy
Trong Nhan Phan	HCMC University of Technology, VNUHCM, Vietnam and FAW, Johannes University Linz, Austria
Thanh Tho Quan	HCMC University of Technology, VNUHCM, Vietnam
Thanh Binh Nguyen	HCMC University of Technology, VNUHCM, and Can Tho University of Technology, Vietnam
Thanh-Van Le	HCMC University of Technology, VNUHCM, Vietnam
Quoc Cuong To	German Research Center for Artificial Intelligence, Germany
Le Minh Sang Tran	Trento University, Italy
Minh-Triet Tran	HCMC University of Science, VNUHCM, Vietnam
Minh Quang Tran	HCMC University of Technology, VNUHCM, Vietnam
Ngoc Thinh Tran	HCMC University of Technology, VNUHCM, Vietnam
Hoang Tam Vo	IBM Research, Australia

Contents

User-Centered Design of Geographic Interactive Applications: From High-Level Specification to Code Generation, from Prototypes to Better Specifications

Christophe Marquesuzaà[1(✉)], Patrick Etcheverry[1], Sébastien Laborie[1], Thierry Nodenot[1], and The Nhan Luong[2]

[1] Université de Pau et des Pays de l'Adour, Laboratoire d'informatique, EA 3000, 64600 Anglet, France
{christophe.marquesuzaa,patrick.etcheverry,
sebastien.laborie,thierry.nodenot}@iutbayonne.univ-pau.fr
[2] Faculty of Computer Science and Engineering, Ho Chi Minh City University of Technology, 268 Ly Thuong Kiet Street, District 10, Ho Chi Minh City, Vietnam
nhan@hcmut.edu.vn

Abstract. This paper deals with models and tools allowing designers of geographic web application to focus on the design work rather than on the related coding problems. The contributions of this paper are a specific design model and its associated design environement. The proposed design model is composed of elements that can be translated, through transformation model technics, into executable source code taking into account high-level specifications of the designers. This automated code generation property offers a design approach based on short cycles where designers may adjust their specifications until the generated application matches their requirements. To facilitate this process, the proposed model has been integrated into a graphical web-based environment allowing designers to visually express their specifications and then to generate and to execute the specified application.

Keywords: Visual design · Geographic web application design · Short lifecycle · Code generation

1 Introduction

Cartography on the Internet has caused a revolution not only in the uses of maps but also in the way to design applications presenting geolocalized data. First research work on geographic information system (GIS) concerned geolocalized data gathering and visualization. A lot of geolocalized data is now available in free geographic databases (geonames.org for example) and new geolocalized

The original version of this paper has been revised: The authors' affiliation has been corrected. The erratum to this chapter is available at https://doi.org/10.1007/978-3-662-54173-9_7

© Springer-Verlag GmbH Germany 2017
A. Hameurlain et al. (Eds.): TLDKS XXXI, LNCS 10140, pp. 1–29, 2017.
DOI: 10.1007/978-3-662-54173-9_1

data can be easily gathered with a simple smartphone integrating a GPS chip. Moreover, simple tools like Google Maps allow any geolocalized data to be displayed in various formats and on several kinds of maps. Consequently, a fair amount of research and development has been conducted on Web-based application generation thanks to Web 2.0 technologies. Particularly in the domain of geographic information system, specific terms appeared in order to designate Internet Geographic Applications [27]: "GeoWeb", "Geospatial Web" and "Web Mapping 2.0". Indeed, many Web-based geographic applications have been developed in different application domains (e.g. tourism, education, surveillance, military) and are using online mapping services (e.g. Google Maps, MapQuest, MultiMap, OpenLayers, Yahoo! Maps or French IGN Geoportail).

Besides information retrieval aspects (such as indexing, query processing or results filtering), this paper focuses on how to design interactive geographic applications, i.e., going beyond a basic search engine interface. Indeed, many application domains require these types of interactive applications: tourism and culture (multimedia encyclopedias...), security (video surveillance...), education (territory discoveries...), etc. A study of these applications led us to identify specific characteristics depending on the application domains but also the following invariants: all these applications handle georeferenced contents, display data on maps or timelines, and offer users interaction possibilities in order to manage, to search, to summarize or to compute geographic data. In this context, implementing geographic applications becomes a difficult task:

– A wide variety of heterogeneous models describe geographic information;
– Different geographic services have to be combined (indexation services);
– Several languages and programming libraries (APIs) have to be exploited.

By taking into account the variability of geographic applications, the diversity of application domains and the complexity of implementation, many works were oriented towards environments facilitating the design and the development of these applications. However, these environments cannot be directly exploited by a designer with no extensive technical skills because they require skills in software development and good knowledge of dedicated technologies (e.g., Javascript, PHP, GIS databases, XML). This conceptual and technical complexity is detailed in the related work (see Sect. 2).

This work particularly focuses on expert designers in their field, such as teachers, who want a rapid prototyping of their (web) applications, i.e., easy to specify, to design, to deploy, to execute and to evaluate their geographic applications. While this type of user is never in a position to formally express all the functional specifications of a geographic application, we assume that an adequate system can allow them to build and to test step by step some specifications by expressing their intentions via adapted graphical abilities. Mashup approaches partially answer this problem and represent a track to explore.

Our contribution to this problem consists in proposing methods and techniques that enable designers to:

– Annotate (semi-automatically) geographic contents owned by an expert. These resources may come from multiple sources (texts, IGN databases, Web...) and must be combined, aggregated and compared according to the designers aims.

- Specify/Evaluate through the graphical abilities how experts may present and emphasize the annotated contents. According to the diversity of application domains (environmental monitoring, tourism, education...), the proposed designed approach considers different facets of the design process that are linked together. Three facets systematically appear in all geographic applications: the "Geographic content" facet (defining the data to integrate and to manage), the "Interface" facet (specifying how data will be displayed in the final generated application) and the "Interaction" facet (defining the possible actions on each data and their behaviours). Of course, other facets may be considered customizing the design process for specific experts.
- Execute the designed applications on several types of devices (on a classical Web browser or on mobile devices). Hence, during the design of an application, the target device characteristics (see the Interface and Interaction facets) and specific device components (GPS, input/output capabilities) have to be considered.

More specifically, the proposed design approach illustrated in Fig. 1 is organized as follows. The designer visually specifies, in his/her preferred order, the contents to be included in the (web) geographic application, how these content are displayed to the end-users and interactive behaviors associated to these contents. This specification is based on generic design models, each model resulting in an instance describing some properties of the final application. All instances are then used by a code generator that automatically computes the executable code of the application. This allows the designer to immediately evaluate his/her specification and to refine it when necessary or to deploy it to target end-users if results are satisfying.

The research efforts that we have been carrying out since 2008 [38] allowed us to achieve significant results validated independently in the following papers:

- Interest of a design process composed of three complementary steps for non-IT specialists [42];
- Evaluation of the content design process step [43];
- Proposal and evaluation of the content design model [39,43];
- Proposal and evaluation of the interaction design process step and its underlying model and visual language [22,40];
- Proposal of the WIND API (http://erozate.iutbayonne.univ-pau.fr/windapi/) [41];
- Proposal and evaluation of the WINDMash (http://erozate.iutbayonne. univ-pau.fr/Nhan/windmash5s/) design environment [38,44].

One goal of this paper is to present how and why the different aspects of our previous works complete each other. Another goal is to show why these works can lead non IT-specialists to smoothly design interactive web applications without having to know anything about the technology background. To address these aims, we propose in this paper to:

- Present a global overview of the design process illustrated in Fig. 1. Previous papers only mentioned the global design process and then focused on a specific design step or model part.

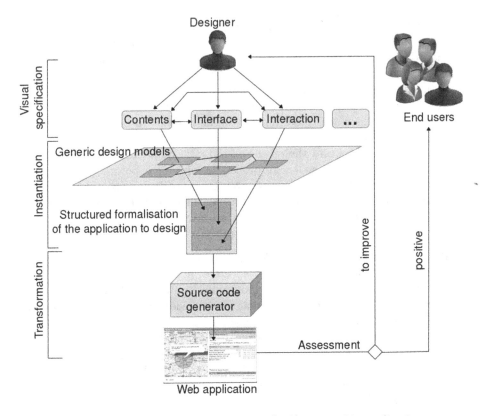

Fig. 1. A design framework dedicated to (web) geographic applications.

– Show the link between the proposed visual design tools of the WINDMash
  demonstrator and their underlying design model.

   The following section details several related works which correspond to the
issues highlighted in this introduction. First, we present the state of the art
of Web environments used by designers for managing and displaying contents,
especially describing Mashup systems. We also give an overview of the current
conceptual models that allow the specification of interactions. Then, we describe
in Sect. 3 our contribution in terms of models referring to the contents and the
user interface of the application (Sect. 3.1) and the model used to define interactions (Sect. 3.2). In Sect. 4, we present an adapted Web-based design environment
named WINDMash. Finally, Sect. 5 presents a conclusion and future work.

2   Related Work

This section presents many systems that enable the development of Web applications. Firstly, several environments facilitating the design of such applications are
described in Sect. 2.1. Existing models for specifying interactions are presented

in Sect. 2.2. Lastly, Sect. 2.3 presents other contributions concerning UML and UML-based models/design methods.

2.1   Web Environments for Managing and Displaying Contents

This section presents some Web environments that help designers to manage and to display data, such as geographic information. We particularly focus on Mashup systems enabling to define contents by combination/aggregation of heterogeneous data with visual tools. These systems are often used to specify the layout of an application with predefined components, like arrays, geographic maps... Mashup systems have been classified according to targeted users (developers, advanced users and end-users).

Table 1 provides a comparison between some existing Mashup systems studied for managing and displaying contents. In addition to the targeted user types, the following criteria have been chosen:

– **Language:** Type of language used to design an application with the Mashup system;
– **Tools:** Tools proposed by the mashup system;
– **Technologies:** Technologies used to implement the mashups;
– **Deployment:** Methods used to embed the mashups in a website;
– **Web services:** Does the Mashup system support Web services;
– **Interactivity:** Does it allow designers to define interactive components.

Summary. Most mashup systems use Javascript for the specification of the Web application interface. Contents are usually defined via data streams but mashup systems are not easy to use by everyone. These systems are generic and are not exclusively designed to develop geographic applications, hence this is the reason why they do not offer a platform for the design of these specific types of applications. Moreover, many mashup systems do not take into account the specification of interactions among the displayed contents in the produced final application.

2.2   Models for Specifying Interactions

This section focuses on models that are used to describe interactions within an application. In the human-computer interaction literature [13,48,56,62], and particularly the work of [1,61], there is no agreed definition of the interaction concept. In this paper, the definitions of [46,65] are merged in order to define an interaction as *"an action usually implemented by a human actor in a system that triggers a reaction that can be direct (visible by the human actor) or indirect (internal to the system, such as a calculation)"*.

Table 1. Comparison of Web environments for managing and displaying contents.

	Language	Tools	Technologies	Deployment	Web services	Interactivity
For developers						
GME[a]	Code	Many	XML	XML code to import in iGoogle	N/A	N/A
Exhibit [30]	Code	Many	HTML, JSON	No	No	Yes but predefined
Chickenfoot [9]	Code	Few	JavaScript	No	No	No
PiggyBank [31]	Code	Few	RDF, JavaScript	No	No	N/A
For advanced users						
Yahoo! Pipes[b]	Data streams	Many	Web standards, YUI	HTML code	Yes	Yes but predefined
Popfly[c]	Data streams	Many	Silverlight	HTML code	Yes	Yes but predefined
Damia [4]	Data streams	Many	REST, XML	No	Yes	Yes but predefined
For end-users						
Afrous[d]	Tools bar	Many	JavaScript	HTML code	Yes	N/A
Marmite [66]	Data streams	Few	Firefox plugin, XUL, JavaScript	N/A	Yes	Yes but predefined
MashMaker [21]	Tools bar	Many	Firefox plugin	N/A	Yes	Yes but predefined
Mashlight [3]	Data streams	Few	JavaScript, XML	No	Yes	Data, Yes but predefined

[a]Google Mashup Editor has been stopped and migrated in Google App Engine.
[b]http://pipes.yahoo.com/pipes/.
[c]Microsoft Popfly has been stopped during summer 2009.
[d]http://www.afrous.com/en/.

Models for Specifying Interactions. In the literature, there are several definitions of the interaction concept. The diversity of the available definitions highlights the complexity of this concept. Furthermore, this complexity is reinforced by the number of models which are necessary to specify interactions [19,34]:

- The *task models* which allow to specify actions that end-users can perform on the system in order to achieve a goal [14]: MAD [59] and its extension MAD* [23], CTT [53], TOOD [45,50], AMBOSS [24], Diane+ [63], UAN [29], GTA [64] or UsiXML [36] which group several models (included a task model) for describing interaction.
- The *dialog models* which are used to describe communications between the user and the system: [11,13,16,25,52].
- The *presentation models* which allow to specify properties and elements that compose a user interface: [26,49].
- The *architectural models* which organize the source code of an interactive application, facilitating its development and its evolution: Seeheim [20,54], AMF [58], Arch [2], PAC [17] or MVC [35].

These different models provide a rich modelling framework that allows to describe the different aspects of interactions. Generally, interaction modelling begins with the formalization of tasks that the user must accomplish. To achieve these tasks, one must define the several dialogs that the system must be able to execute. Then, presentation models enable end-users to visually accomplish their tasks. Architectural models, at the end, provide some source codes that comply with the properties specified in the above mentioned models.

However, using these different models requires some abstraction and modelling capabilities that are specific to specialists in computer science. So designers must be able to clearly separate the different levels of abstraction that can describe the different layers of an interactive system.

In the GIS context (ArcGIS or IGN for instance), the data is ordered according to specific layers: rivers, transport networks, towns, etc. This organization into layers allows designers to define specific interactions, such as displaying specific attributes and the way they are presented: changing the appearance of the layer, placing the layer in the background, etc.

In this work, it is important to preserve this layered organization and to allow designers to define interactions over geographic data, taking into account the layer on which they belong to. This implies to be able to semantically type each data and to allow designers to specify interactions on this typed data, e.g., highlight all the data with the type "rivers". Moreover, the targeted applications should be designed by non-technical designers without any help from computer scientists and should address at most one or two tasks.

These two reasons lead us to avoid any modelling framework that would be too complex:

- The task models seem too heavy to use regarding the size of the considered applications to produce and it does not fit our targeted designers capabilities.
- The dialog models seem unavoidable as they describe the possible user actions on a system as well as the system reactions. This kind of model is the core of an interaction and our proposal must provide some techniques for expressing such dimension of the interaction.
- The presentation models also cover an important aspect of the interaction as they describe the visible layer of an interaction. Due to its visual nature, this model is particularly important for non-expert designers who want to graphically design the interactions.
- The architectural models play a secondary role for our targeted designers. In order to develop tools for automatic code generation, developers should not need to worry about the way to organize the final code of their applications.

According to the models mentioned above, our proposal is at the crossroads of dialog models and presentation models. The goal is to allow the designer to describe the elements of the interface (presentation layer) with which the user can interact. In the context of geographic applications, the designer will specify which geographic information has to be presented on the interface and where the user can interact.

This specification will conduct the designer to describe the interactive features of the application based on the displayed data. These interactive features will specify possible dialogues between the user and the system (dialog layer).

We therefore propose a hybrid approach that allows non-IT designers to describe the elements that can be rendered on an interactive interface and the consequences of a user action on the displayed items.

2.3 Particular Case of UML and Derivatives

UML offers a standard modelling to specify the different facets of an application. Since version 2, UML offers thirteen specialized models allowing the designer to describe the various features of an application. These diagrams can be viewed as visual languages. These languages offer graphical notations to a designer which let him/her specify the application he/she wants to design. Many CASE tools, such as IBM Rational Rose (www.ibm.com/software/fr/rational), Modelio (www.modeliosoft.com), BOUML (www.bouml.fr), and research works have shown that it is possible to generate source code from a UML specification. As shown in [7,33,67], UML can be used as a modelling language to generate Java code from class diagrams and state transition diagrams. Interactive aspects of the system are described by state transition diagrams that specify the interactive possibilities of the user depending on the state of the system. It can also describe the internal reactions of the system after a user action. The work presented in [28] highlights the interesting capabilities of UML sequence diagrams to describe on-the-fly, at runtime, the user interaction with his/her application.

Sequence diagrams describe the interaction with flows exchanged between the user and the system, and also between the system components. With sequence diagrams, it is possible to ignore the devices used by the user and highlight, with messages, what the user can do on the system regardless of the input device used. These diagrams are also the only ones to include a representation of the user and its role in the interaction. The representation of the interaction as a stream describing the interaction between the user and the system appears as a natural way to describe what happens between a user and a system during an interaction. Unlike the state-transition diagrams, they allow the designer to describe the interactive dimension of the system using several steps and not globally or for a single class. This modular property of the interaction also seems interesting for non-IT designers that have difficulties to consider the interactive dimension of an application in its globality, especially if the designed system offers many interactive facilities.

WebML [15] is a language for designing traditional data-intensive Web applications. It has a set of visual notations to model the content, structure and navigational aspects of a Web application. It has a visual formalism to model relevant interaction and navigation operations. The design of a Web application is based on two orthogonal perspectives: data (for instance, describe with ER data model) and navigation (specified by the WebML Hypertext Model). It has an extension dedicated to GIS: WebGIS. Some approaches such as [6,10,18] propose to automatically generate WebGIS applications able to recommend content to users and adapt the interface of a Web-based spatial map. The approach is based on an extension of WebML visual formalism and the supporting desktop CASE tool WebRatio. WebGIS is adapted to elaborate geographic applications integrating maps but does not propose specific artifacts to handle other geographic features such as calendars or timelines.

A context-aware meta-model for adapting a Web GIS interface in the context of emergency scenarios has been proposed in [5]. This model is based on the UML and the MDA (model driven architecture) paradigm. More precisely, they automatically transform the displayed data to the capabilities of the end-user devices and their preferences or needs. For that purpose, the authors propose a full-Web framework for generating adaptive GIS applications to a specific context but they do not provide beforehand visual editing tools dedicated to non-IT specialists as well as a flexible design process.

There are other methods and tools for the design and development of web application such as UWE [32], UWA [60], OOH4RIA [47], ActionGUI [8] or MontiWIS [55]. Most of them follow the principle of "separation of concerns" using separate models for different views on the application, such as content, navigation, presentation and business processes. Most of them propose extensions for code generation: WebRatio for WebML, MagicUWE and ActionUWE [12] for UWE. However, most of these code generation tools are not dedicated to geographic applications, and are only available as traditional software solutions and not as full-Web tools.

3   Models and Visual Language for Designing Web Geographic Applications

Related work presented in Sect. 2 highlights a set of models allowing designers to build geographic Web applications. These models were not thought to be easily coupled. This implies either an additional inter-connected work for the designer or a partial application modelling focused on one of the models at the expense of the others. Based on that observation, we propose a global model summarized in Fig. 2. This model is composed of three inter-connected submodels linked by the Annotation concept. This central concept can be enriched with semantic information defined in external knowledge bases such as ontologies or linked open data entities. This semantic enrichment benefits to the three model dimensions. It allows designers to specify, in a complementary way, various geographic data (Content model - see Sect. 3.1), the layouts for each data (Interface model - see Sect. 3.1) and also associated interactive behaviours (Interaction model - see Sect. 3.2). With the suggested model, annotated contents remain a central focus of the design approach: designers define contents, display them on the GUI and make them interactive.

3.1   Content and Interface Models

This design step focuses on the (geographic) contents that should be integrated in the application. Contents represent the central concept of our design approach which aims at enhancing specific geographic data defined by designers. This enhancement is carried out by:

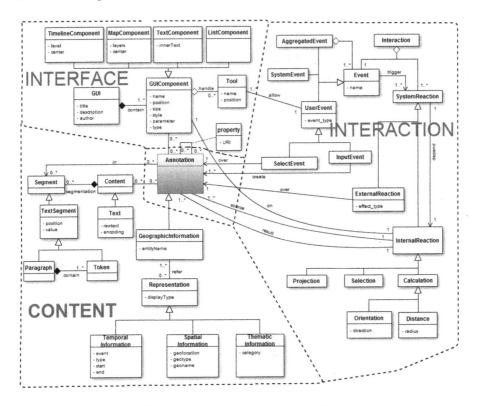

Fig. 2. Global model for the design of geographic applications

- a presentation of these contents (on the screen) according to several representation modes;
- an enhancement of these contents through specific user interactions.

Content Model. We propose a model (Fig. 3) to specify contents of geographic Web applications [43]. Contents (**Content**) ② can be composed of several segments (**Segment**) ③. As illustrated in Fig. 3, contents can be composed of several annotations (**Annotation**) ① referring to specific segments. An annotation can be connected with other annotations with different properties (**property**) ④. For example, an annotation about *"Liverpool"* is related to an annotation referring to *"Manchester"* by a property whose URI is `wind:nextTo`. Of course, thanks to URI attribute of the **property** class, both annotations can be also linked to external semantic concepts, e.g., specifying that *"Liverpool"* and *"Manchester"* are *"cities"* located in *"England"*. This relationship can be exploited later when filtering contents and/or when defining interactions on these annotations.

Currently, the considered services extract geographic contents from textual sources. Consequently, as presented in Fig. 3, texts (**Text**) ⑤ can be segmented into paragraphs (**Paragraph**) ⑥ and tokens (**Token**) ⑦ that inherit from

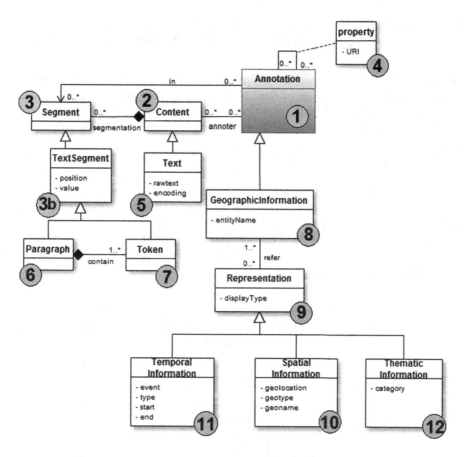

Fig. 3. Model used to describe geographic contents

a textual segment (TextSegment) ③b. A textual segment can have one position (for example, 6th word, 4th paragraph) and a value presented as a string (for example, "Paris", "I live 10 kilometers from the south of Paris.").

In addition, geographic information (GeographicInformation) ⑧ is a kind of (inheritance relationship) annotation ① that can have several representations (Representation) ⑨ related to the spatial dimension (Spatial-Information) ⑩, the temporal dimension (TemporalInformation) ⑪ or the thematic dimension (ThematicInformation) ⑫.

Spatial representation of an annotation always results in geographic coordinates to locate the annotation on a map. This representation is described according to a form that can be a point (for example a city), a line (for example a river or a road) or a polygon (for example a region). To conform with traditional GIS (see Sect. 2.2), each spatial information must be classified to allow designers to define interactions on the layer to which this information belongs.

Temporal representation of an annotation can be a specific moment (for example "*2013-08-16*") or a period (for example "*From September 1940 to April 1942*").

Interface Model. This design step allows designers to organize the interface of their application in terms of size, position, map provider, zoom level... An interface is composed of displayers that display geographic information previously defined at the *Content* design step. Designers can also decide where and how each displayer is presented on the screen.

This section aims at presenting the model elaborated to present geographic contents on the graphic interface of the final application. The application interface is considered as a visualization layer to present contents under various forms. An interface is built by assembling several interface components, each component being specialized to represent contents in a specific form.

Even if the *Interface* model is not a strong contribution from a research point of view, this model (Fig. 4) is necessary to display geographic information defined in the previous design step.

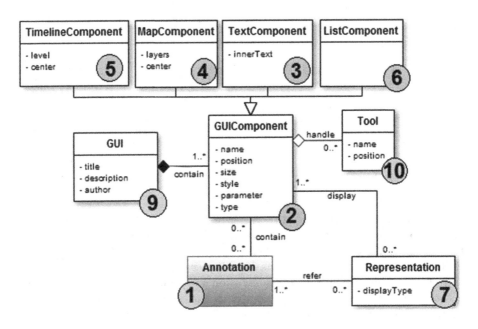

Fig. 4. Model used to describe user interfaces

A geographic application integrates a graphic user interface (GUI) ⑨. This interface can be composed of several graphic components (GUIComponent) ②️ organized according to a specific layout. Currently, four types of interface components are considered: text (TextComponent) ③, maps (MapComponent) ④, timelines (TimelineComponent) ⑤ and lists (ListComponent) ⑥ which are subclasses of `GUIComponent` (2) in Fig. 4. Each user interface component supports its own displayed parameters. An annotation (1) can appear in one or several interface components (2). The displaying form of an annotation depends on the interface component type:

- For a `TextComponent` (3), geographic contents are represented in a textual form and most of the time with the name of a geographic place/area.
- For a `MapComponent` (4), geographic contents are displayed as geometries on a map. A point can represent a place, a line can represent a road, a river or an itinerary for example. A polygon can represent a region, a city delimitation... Designers can choose the cartographic background for their application. The `MapComponent` can support several maps providers, such as GoogleMaps, french IGN Geoportail.
- The `TimelineComponent` (5) displays temporal information (for example, dates, periods) on a chronological line, such as the SIMILE timeline (www.simile-widgets.org/timeline/).
- The `ListComponent` (6) displays annotations in a bulleted list.

Each interface component can have one or more associated tools (`Tool`) (10) allowing the final user to create a new annotation when interacting with this interface component: for example, a pencil to manually annotate a text in a textual component or buttons to draw points, lines and polygons on a map component.

When an annotation (7) is displayed on an interface component, designers may have to specify the representation of this annotation. For example, an annotation related to "*Bayonne*" can have several representations on a map component:

- a cartographic representation as a point whose longitude is -1,475 and latitude is 43.4936, its geolocation is thus `POINT(-1.475 43.4936)` ;
- another cartographic representation as a polygon `MULTIPOLYGON(((-1.49222 43.50884, -1.49223 43.50901, ..., -1.49222 43.50884)))`.

3.2   A Model for Interaction Design

This section describes a model facilitating the design and the implementation of interactions on geographic contents presented on interface components.

An Interaction Model Focused on Contents. This section proposes an interaction model built upon the following assumption: a final application displays contents on an interface and users can interact with these contents. The result of an interaction is the enhancement of existing contents or the creation of new contents.

According to this approach, interaction design is guided by (geographic) contents that must be enhanced. Content enhancement can be carried out by user interactions on displayed contents.

The proposed interaction model is described in Fig. 5. An interaction ③ is defined as a specific event ④ triggering a system reaction ⑩. The triggered events can be sent by the system ⑦ or created by a user action ⑤. Triggered events can also be an aggregation ⑥ of system and/or user events (for example, it is 10:00 a.m. and the user carried out specific actions).

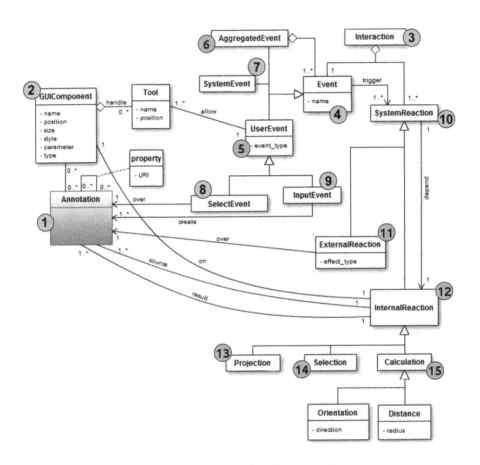

Fig. 5. Model used to describe interactions

A user event ⑤ can be defined according to two kinds of actions: a selection action or an input action. A selection action ⑧ is defined as the selection of an annotation ① displayed on the interface ② while an input action ⑨ is characterized by an annotation creation ① in an interface component ②. In this model, user actions only deal with selecting (with a click, a mouseover, a drag and drop...) annotations displayed on the screen or creating new annotations with specific tools displayed on the interface (pencil annotation on a textual component, drawing points, lines or polygons on a map component...).

System reactions (10) can be external or internal. External reactions (11) are system reactions that are visually displayed to the user. These reactions are defined by a visual effect (show, hide, highlight, zoom...) applied on an annotation (1) displayed on the interface (`GUIComponent`) (2). Internal reactions (12) correspond to operations that will create a new annotation, modify or move an existing annotation. The current model considers three kinds of internal reactions:

- Projection operations (13) copy an existing annotation (1) from an interface component (2) to another;
- Selection operations (14) identify which annotation (1) was selected by the user among all annotations displayed on an interface component (2);
- Calculation operations (15) aims at computing new annotations starting from existing annotations or previously calculated annotations. Within geographic applications, these operations will be of spatial or temporal nature, e.g., determining the country to which belongs a city, calculating the number of days between two dates.

For any interaction, a user action can generate one or more internal reactions and always one or more external reactions that will apply visual effects on one or more annotations. The user will visually notice these visual feedbacks.

4 Our Environment and Tools for Generating Web Geographic Applications

To define the features of a design approach allowing inexperienced designers to define (web) geographic applications without any computer scientist help, we focused on traditional software engineering approaches, on development paradigms dedicated to non-experts and on tools allowing to generate applications without any specific technical skills, more specifically:

- "agile" design approaches which recommend to work with short lifecycles involving end-users.
- programming visual languages because of their graphical expressiveness allowing their handling by non-experts.
- Mashup tools allowing to visually implement small applications without any programming skills.

The functional features of the design environment must include the following properties:

- a set of dedicated services to define the contents to be integrated in the application. These contents may be extracted from geographic databases or from geographic texts (travel books, topoguides...). The environment must also provide services allowing designers to refine and to combine any extracted contents in order to create new contents semantically interesting for the designer.

– a set of items allowing to build graphical user interfaces displaying the geographic contents. According to the tripartite nature of these contents (see Sect. 3.1), these interface components must be able to display the spatial, the temporal and the thematic dimension of any geographic content. Hence, it should be easy for them to create applications with dedicated interfaces: maps for the spatial dimension of the contents, calendars or timelines for the temporal dimension and textfields for any dimension.
– graphical abilities allowing designers to specify the detail of any user interaction with displayed contents.
– the design process should be flexible with different starting points. For example, some designers could first define the contents to emphasize, next they could decide how to display them on a GUI and finally they could specify the associated interactions. This flexibility allows to partially take into account empirical design approaches used by non computer scientists designers.
– the code of the designed application must be automatically generated. This property supports short lifecycles and gives designers immediate feedbacks of their expressed requirements.

To ensure the executability of the *Content, Interface* and *Interaction* models, respectively detailed in the Sects. 3.1 and 3.2, and to overcome the development complexity associated with the used technologies, a dedicated Javascript API called WIND (*"Web INteraction Design"*) [41] and the WINDMash associated environment have been implemented.

The authoring-tool called WINDMash is composed of three modules corresponding to the three steps of the design process:

1. A pipes editor which allows to combine different services and to filter the geographic contents handled by the application (*Content* phase);
2. A graphical layout editor which is used to display, for instance, mapping and/or timeline components and/or multimedia contents (*Interface* phase);
3. A UML-like sequence diagram builder which allows to specify potential end-user interactions on the displayed contents (*Interaction* phase).

WINDMash is a design environment integrating the proposed unified design model (see Fig. 2). It offers dedicated tools allowing designers to handle graphically each part of the model in order to create a specific (web) geographic application. Each module creates instances (RDF/XML descriptions) of a specific part of the model. When designers want to check their specification, all instances are merged to automatically generate the executable code of the application. The generated code is based on the Javascript WIND API (http://erozate. iutbayonne.univ-pau.fr/windapi/) which complies with the proposed model.

4.1 Content Phase

In order to manage the geographic contents (contents and annotations), a mashup editor has been developed. It allows designers to create a processing chain with different services aiming at extracting and/or managing geographic

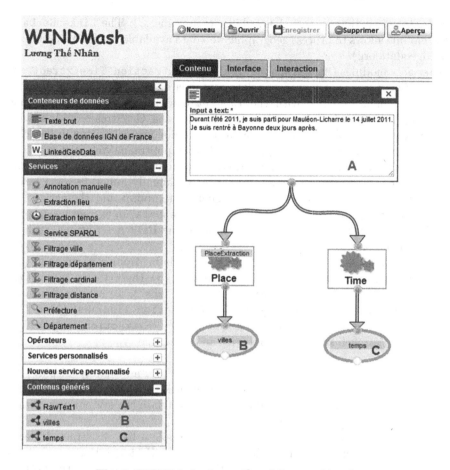

Fig. 6. WINDMash pipes editor (*Content* Phase)

contents (Fig. 6). This tool is inspired from Yahoo! Pipes (http://pipes.yahoo.com) editor.

Starting for example from one or several raw texts, designers may easily define a workflow by using WINDMash dedicated services. This activity automatically transforms an entry text into a structured document in accordance with the *Content* model. These contents may be later displayed (*Interface* phase) thanks to dedicated displayers: `TextDisplayer`, `MapDisplayer`, `TimelineDisplayer` and/or `ListDisplayer`. These modules may be set up by designers according to their specific goals. There are two groups of modules: `DataContainers` (*"Conteneurs de Données"*) and `Services`.

A *data container* may be a raw text (`Texte brut` - `Raw text`), a geographic database (`French IGN`) or any geographic web data (`LinkedGeoData`). A raw text is a textual document (for example a story tale) used by the designer and linked to a `Service` module. An IGN database has geographic information tables (cities, towns, regions, rivers, lakes, hills, mountains...). The `LinkedGeoData` data module allows to process geographic web data available on the Web (http:// linkedgeodata.org).

Services modules (`ManualAnnotation`, `PlaceExtraction`, `TimeExtraction`, `SPARQL Service`...) may process a data container and generate contents (structured data in accordance with the *Content* model detailed in the previous section). The `PlaceExtraction` module invokes the GeoStream Web service [57] to extract spatial information from a textual document (written in French). The `TimeExtraction` module implements the TempoStream Web service [37] to extract temporal information from a textual document. From a given text, the `ManualAnnotation` module allows designers to manually annotate some terms according to the three following facets: spatial, temporal and thematic. The `SPARQL Service` is used with Web data (`LinkedGeoData`) to extract geographic entities using a SPARQL request applied on online RDF descriptions, such as Linked Open Data.

Figure 6 illustrates the following processing chain: from a given text (A), designers want to automatically extract places (B) (here towns and cities: "`villes`") and temporal references (C) (called "`temps`"). To achieve this goal, they call the place extraction (web) service (`Extraction lieu`) detailed in [57]. They also call the temporal extraction web service (`Extraction temps`). The place extraction web service may identify and mark the type of extracted entities.

Once the workflow is defined, it is possible to immediately display the computed geographic contents with a double-click on the generated contents "`villes`" (B) and "`temps`" (C). Designers may validate the system proposition or delete some inappropriate annotations.

Whenever contents are computed, as for example a list of extracted places (B), WINDMash automatically generates a RDF/XML description corresponding to the *Content* phase. These descriptions are also available on the bottom left of the WINDMash prototype, in the "`Generated Content`" ("*Contenus générés*") area (Fig. 6).

Next paragraph shows that these descriptions may also be used in the WINDMash graphical layout editor to display these contents inside dedicated displayers.

4.2   Interface Phase

The graphical layout editor allows designers specifying the GUI of their geographic web application. Designers may define which type of displayers they want: for example, a text displayer (`Afficheur texte`), a map displayer (`Afficheur carte`), a list displayer (`Afficheur liste`) or a timeline displayer (`Afficheur frise`). Designers also define how these displayers are presented (title, size, position...).

Figure 7 shows how designers may specify the graphical layout of their applications. The left menu presents three separate parts:

– the various types of available displayers (`Afficheurs` on the top);
– all the available contents (`Contenus générés` on the middle) previously computed by the workflow editor (see Sect. 4.1);
– the displayers which are currently used (`Afficheurs générés` on the bottom).

As presented in the *Content* phase, the different displayers and generated contents which are extracted from the *Content* phase may be used by designers with drag-and-drop operations. Designers may define the precise size and position of any displayer. They may also define the name of a displayer and this name will be later used in the generated displayers (*Afficheurs générés*) area. This name will also be used in the *Interaction* phase (see Sect. 4.3). In Fig. 7, three displayers have been specified: a text displayer ① (`Texte`), a map displayer ② (`Carte`) and a timeline displayer ③ (`Frise`).

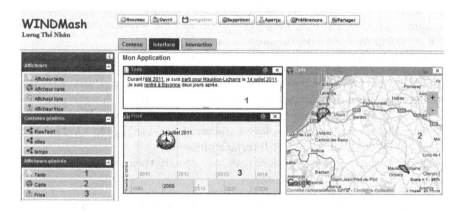

Fig. 7. WINDMash graphical layout editor (*Interface* Phase)

When designers drag-and-drop any textual displayer (Text or List Displayer) into the design area, this displayer is empty[1]. To display any content inside a displayer, designers must drag and drop the content from the generated contents area. For example, if designers want to display the text written in Fig. 6 into the textual displayer ①, they must drag the `RawText1` content from the menu and drop it inside the text displayer called `Texte`. Next, to automatically underline the places and the temporal information extracted from this text, designers must both drag the `villes` (towns and spatial references) and `temps` (temporal references) contents and drop them inside the text displayer called `Texte`.

Similarly, with other displayers, if designers want to display places (resp. temporal references) on a map (resp. timeline), they must drag the `villes` (resp. `temps`) contents and drop them on the map (resp. timeline) displayer called `Carte` (resp. `Frise`). It is also possible to set up the style of a text (size, font...)

---

[1] The map displayer displays a default map and the timeline displayer is centered on the current date.

inside a text displayer, or the map provider (Google Maps or Yahoo! Maps or IGN Geoportail...) inside a map displayer.

Up to this step, designers may use WINDMash to automatically generate two instances of the *Content* and of the *Interface* model part. They may preview a (static) web geographic application by clicking on the preview (Aperçu) button located on the top of the menu bar. They may also save and modify the application at any time. The generated application remain "static" because it does not have any dynamic behaviour due to the end-users actions (click or mouseover for example). This dynamic aspect is considered in the next section.

4.3   Interaction Phase

To make the design of interactions easier, we have already proposed in [40] a visual language which has been integrated in the WINDMash environment. The example presented in Fig. 8 has the following specification: *"when users left click on a place quoted in the text, the system identifies the selected place, highlights it and transfers this place on the map to zoom-in."*

The design workspace is divided into two areas:

- a bottom left area (Interaction components - Blocs d'interaction) presents the set of five interaction components available for designers in order to define interactions from the five interaction components. This area also presents the set of interface displayers (Generated Displayers - Afficheurs générés) created/set up in the Interface phase;
- a central working area where designers may specify their desired interactions with the proposed visual language.

Similarly to the two other working spaces of the WINDMash environment, the specification of an interaction is graphically defined with drag-and-drop operations combining the different components of the visual language. During the specification of a user action or of an internal or external action on an interface component, the WINDMash environment automatically proposes to the designer the whole list of annotations available on this component[2].

The system also requires the designer to precisely define the set of annotations involved in the current interaction. According to the interaction specification expected by the designer, the designer successively drags-and-drops the available components proposed into the left menu/area and parameterizes them. The designer may specify that the current interaction concerns any spatial annotation, or all annotations with a given type (here "Town"), or a specific annotation among those previously defined.

As in the previous design steps, the graphic specification of an interaction is automatically translated into an RDF/XML description that instantiates the model presented in Fig. 5. This RDF/XML description of the interactive abilities can be considered as equivalent to the structured requirements defining the behaviour of the system to run. This specification may be merged with the two

---

[2] This is a first level of guidance for the designer.

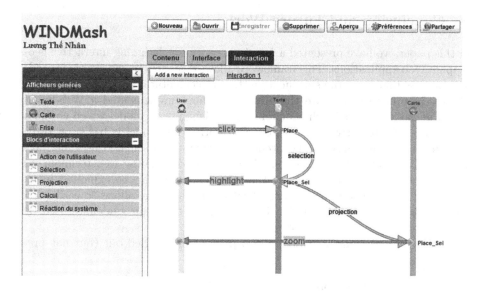

Fig. 8. Graphical specification of an interaction with the WINDMash environment

other RDF/XML descriptions extracted from the *Interface* phase and from the *Content* phase. The complete RDF/XML files which describe the application example presented in this section can be found at the following URL: http://goo.gl/fBYfaa. All RDF/XML files refer to the concepts defined in the global model presented in Fig. 2. To generate an executable program corresponding to the whole visual specification of the three phases, the WINDMash environment exploits the different merged RDF/XML descriptions thanks to the SPARQL query language (http://www.w3.org/TR/rdf-sparql-query/). Thereafter, predefined rules are used to generate Javascript objects invoking WIND API functionalities. The Javascript generated code, which is interpreted by the web browser, is thus very synthetic. Thanks to this code generation feature, designers may run their application and have an immediate feedback on the specified interactions. As presented in Sect. 1 and Fig. 1, if designers are not fully satisfied, they may come back to the visual design phase and modify the interaction diagrams.

The interaction model focuses on simple interactions and not on a set of interactions supporting complex user tasks. Currently, considered events are limited to system events (click, on focus, highlight...) and not all types of discrete events. Aggregated events (combinations of events) specified in the model are not yet implemented in the current version of the WINDMash prototype. These choices were made in order to facilitate design activities for non computer-scientists. As shown by our experiments [22,40], the design of interactions is the more complicated step for non-computer-scientists. We have decided to conceive an interaction model adapted to non-expert designers in order to foster their design experience.

5   Conclusion and Future Work

In this paper, we have presented a new framework for designing interactive geographic Web applications. This framework allows designers to express their needs in a process that links the specification, the implementation and the evaluation phases. In order to facilitate this design, we have implemented a Web-based platform named WINDMash that relies on a rich generic model which covers a wide range of interactive applications. Currently, this model is based on three fundamental design facets: the content to manage, the associated user interface and the specification of interactions. We have used semantic web technologies to encode this model and to show that the semantics can be exploited in all the design facets. More precisely, the RDF/XML language allows to easily create, enrich, connect all the annotations which are the main guidelines used by designers to elaborate their applications.

A set of evaluations focusing on two points were carried out (but not presented in this paper):

1. The ability of the proposed model and visual language to specify and to implement interactions based on geographic contents.
2. The ability to easily design and to implement geographic applications corresponding to a specific type of designers: teachers and their pedagogical requirements.

Each one of these evaluations was based on a test protocol with several steps and their corresponding evaluation criteria [38].

The first evaluation, published in [40], focused on the evaluation of the visual language and the underlying model allowing designers to easily and simply specify the interactive behaviour of a geographic application.

The second evaluation deals with measuring the abilities and limits of the WINDMash environment to design educative applications dedicated to relevant uses for elementary school. In this experimentation, teachers produced hand-made mockups and we measured the capacities and limits of WINDMash to meet this need. In her Master thesis, a complex scenario mixing simple interactions and correlation annotations has been defined in [51]. We proved that the WINDMash environment allowed to fully design and to fully implement this scenario [38].

Currently, we work on a third evaluation concerning the complete design process and more precisely on the ability offered to non computer-scientists to design geographic applications by themselves without any help. We work on defining the complete test protocol in order to evaluate:

– the relevance of a three facets design process for designers that are not computer-scientists;
– the flexibility of the process in terms of back and forth between each design step.

Some elements of this evaluation were implemented in [22]. First results show that for most of the testers (non computer scientists but ICT aware):

1. the current 3 phases process (*Content* ↔ *Interface* ↔ *Interaction*) made the gradual improvement of the application design easier.
2. the experimentation allowed a good general evaluation of the WINDMash environment. Non-computer scientists will be able to use it when the identified coding bugs will be fixed.

Furthermore, we wondered about the methodological aspect of the design process and, specially, about how chaining the steps to specify the data, the interface and the interactions while going back and forth. The goal was to evaluate if it was possible to identify natural sequences in order to highlight reusable methodological patterns.

In our case study we have identified three scenarios:

- Scenario 1: Firstly, designers specify all the geographic data that have to be handled by the application, i.e., the locations and the periods cited in the initial text. Then, they build the graphical user interface by inserting two textual components (one for displaying the initial text and another one for displaying the "department" names), two maps components to highlight locations, and one timeline component to display dates and periods. Finally, designers define all the possible user interactions related to the data displayed on the graphical interface.
- Scenario 2: Designers build first the graphical user interface by inserting the graphical components (i.e., text, map and timeline). Since no geographic data has been previously defined, these components are either empty (such as the textual components) or contain a default presentation (such as a world map or a timeline centered on the current day). Then, they define from the initial text, all the required geographic data that must be integrated/displayed in their application. Then, they come back to the graphical layout design in order to insert all the data. Finally, designers define all the possible user interactions.
- Scenario 3: From the initial text, designers specify a first subset of geographic data (e.g., they first focus exclusively on locations and not on the temporal information). Then, they build a first version of the graphical user interface, (e.g., beginning with the textual components). Then they go back to the previous step in order to refine the geographic data (e.g., computing the "department" names from the locations quoted in the text). Thereafter, they complete the GUI by adding the map components. Next, they specify the user interactions associated to these elements. Then, they execute and evaluate this preliminary version of their application. Afterwards, they decide to update their application by (a) defining the useful temporal information, (b) adding the timeline component with these information and (c) specifying the associated user interactions. Designers have switched back and forth between these three steps all along the design process.

Several points can be highlighted after this experiment:

- Scenario 1 has been followed by very few designers that had sufficient skills to define at once all the required geographic data that have to be integrated in the graphical user interface.

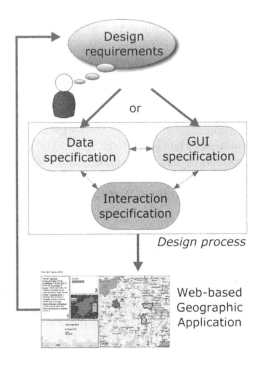

Fig. 9. Observed design processes

- Scenario 2 has been followed by some designers that remembered that it was possible to initially build the graphical user interface before computing the required geographic data.
- Scenario 3 has been followed by most designers who did not succeed in identifying in one step all the necessary data. Thus, they had to alternate several times between the data and the interface design steps.

We have observed (Fig. 9) that most designers would have preferred to start designing the graphic interface of their application because this dimension is most visual and allows them to evaluate concretely if their design choices match with the final application. The possibility to alternate between design phases remains essential to allow designers building their application step by step, in an iterative and incremental way, without carrying out heavy tiresome preparation work.

Several extensions of the existing facets can also be considered. Currently, the content facet is limited to the processing of textual sources with spatial and temporal indexation services. Considering multimedia contents, such as videos, images or sounds, would firstly allow us to manage a wide variety of geographic sources and also to integrate other types of multimedia indexing services. In addition, we will be able to produce a large palette of Web applications with other specific interactions on media. As illustrated in Fig. 1, new facets can be considered in our approach. For example, a user facet will allow designers to customize the generated applications with contents and interactions related to their preferences.

Our model is thought to support the changes previously cited. Indeed, we have already conducted some experimentations, not described in this paper, on the following points:

– Concerning the exploitation of multimedia contents, we have produced examples of Web applications using the Timesheets library (http://tyrex.inria.fr/timesheets/). This library allows to synchronize within a Web page several multimedia contents, some data on maps and/or timelines. It takes into account the user actions on these elements.
– On the user side, our prototype has already the ability to focus on geographic information based on a given reference, e.g., display all the cities within a 20 Km area. We plan to adapt generated applications by customizing displayed contents and available interactions basing on users preferences and location.

Based on these experiments, we found that our design model is extensible and already goes beyond the results presented in this paper. In the future, we want to improve our code generator to easily integrate new extensions of our model and to produce other types of targeted applications, such as native applications for mobile devices.

References

1. Hewett, T.T., Baecker, R., Card, S., Carey, T., Gasen, J., Mantei, M., Perlman, G., Strong, G., Verplank, W.: ACM SIGCHI Curricula for Human-computer Interaction. Technical report. ACM, New York (1992). ISBN:0-89791-474-0. http://dl.acm.org/citation.cfm?id=2594128
2. Bass, L., Faneuf, R., Little, R., Mayer, N., Pellegrino, B., Reed, S., Seacord, R., Sheppard, S., Szczur, M.R.: A metamodel for the runtime architecture of an interactive system: the UIMS tool developers workshop. SIGCHI Bull. **24**(1), 32–37 (1992). ACM. ISSN:0736-6906. http://dl.acm.org/citation.cfm?id=142401
3. Albinola, M., Baresi, L., Carcano, M., Guinea, S.: Mashlight: a lightweight mashup framework for everyone. In: 2nd Workshop on Mashups, Enterprise Mashups and Lightweight Composition on the Web (2009)
4. Altinel, M., Brown, P., Cline, S., Kartha, R., Louie, E., Markl, V., Mau, L., Ng, Y.-H., Simmen, D., Singh, A.: Damia - a data mashup fabric for intranet applications. In: Proceedings of the 33rd International Conference on Very Large Data Bases, pp. 1370–1373 (2007)
5. Angelaccio, M., Krek, A., D'Ambrogio, A.: A model-driven approach for designing adaptive web GIS interfaces. In: Popovich, V., Claramunt, C., Schrenk, M., Korolenko, K. (eds.) Information Fusion and Geographic Information Systems. Lecture Notes in Geoinformation and Cartography, pp. 137–148. Springer, Heidelberg (2009)
6. Avagliano, G., Di Martino, S., Ferrucci, F., Paolino, L., Sebillo, M., Tortora, G., Vitiello, G.: Embedding google maps APIs into WebRatio for the automatic generation of web GIS applications. In: Sebillo, M., Vitiello, G., Schaefer, G. (eds.) VISUAL 2008. LNCS, vol. 5188, pp. 259–270. Springer, Heidelberg (2008)

7. Barbier, F.: Supporting the UML state machine diagrams at runtime. In: Schiefer-decker, I., Hartman, A. (eds.) ECMDA-FA 2008. LNCS, vol. 5095, pp. 338–348. Springer, Heidelberg (2008)
8. Basin, D.A., Clavel, M., Egea, M., de Dios, M.A.G., Dania, C.: A model-driven methodology for developing secure data-management applications. IEEE Trans. Softw. Eng. **40**(4), 324–337 (2014)
9. Bolin, M., Webber, M., Rha, P., Wilson, T., Miller, R.C.: Automation and cus-tomization of rendered web pages. In: Proceedings of the ACM Conference on User Interface Software and Technology (UIST), pp. 163–172 (2005)
10. Brambilla, M., Fraternali, P.: Large-scale model-driven engineering of web user interaction: the WebML and WebRatio experience. Sci. Comput. Program. **89**(Part B), 71–87 (2014). Special Issue on Success Stories in Model Driven Engineering
11. Britts, S.: Dialog management in interactive systems: a comparative survey. Sci. Comput. Program. **18**(3), 30–42 (1987)
12. Busch, M., Ángel García de Dios, M.: ActionUWE: transformation of UWE to ActionGUI models. Technical report, Ludwig-Maximilians-Universität München, Partners: LMU and IMDEA. NESSoS, Project (2012)
13. Caffiau, S., Girard, P., Guittet, L., Scapin, D.L.: Hierarchical structure: a step for jointly designing interactive software dialog and task model. In: Jacko, J.A. (ed.) HCI 2009. LNCS, vol. 5611, pp. 664–673. Springer, Heidelberg (2009). doi:10.1007/978-3-642-02577-8_73
14. Card, S.K., Moran, T.P., Newell, A.: The Psychology of Human-Computer Inter-action, 1st edn. CRC Press, Boca Raton (1983)
15. Ceri, S., Fraternali, P., Bongio, A.: Web Modeling Language (WebML): a modeling language for designing web sites. Computer Networks **33**(1–6), 137–157 (2000)
16. Churcher, G.E., Atwell, E.S., Souter, C.: Dialogue management systems: a survey and overview. Technical report, School of Computer Science, University of Leeds (1997)
17. Coutaz, J.: PAC, on object oriented model for dialog design. In: Interact 1987, pp. 431–436 (1987)
18. Di Martino, S., Ferrucci, F., McArdle, G., Petillo, G.: Automatic generation of an adaptive WebGIS. In: Carswell, J., Fotheringham, A., McArdle, G. (eds.) W2GIS 2009. LNCS, vol. 5886, pp. 171–186. Springer, Heidelberg (2009)
19. Diaper, D., Stanton, N.: The Handbook of Task Analysis for Human-Computer Interaction. Lawrence Erlbaum, Mahwah (2004)
20. Dragicevic, P., Fekete, J.-D.: Support for input adaptability in the ICON toolkit. In: Proceedings of the 6th International Conference on Multimodal Interfaces, ICMI 2004, pp. 212–219. ACM, New York (2004)
21. Ennals, R., Garofalakis, M.: Mashmaker: mashups for the masses. In: Proceedings of the 27th ACM SIGMOD International Conference on Management of Data, pp. 1116–1118 (2007)
22. Etcheverry, P., Laborie, S., Marquesuzaà, C., Nodenot, T., Luong, T.N.: Con-ception d'applications web géographiques guidée par les contenus et les usages: cadre méthodologique et opérationnalisation avec l'environnement WINDMash. J. d'Interaction Personne-Système **3**(1), 1–42 (2015). https://hal.archives-ouvertes.fr/hal-01162916v1
23. Gamboa, F., Scapin, D.: Editing MAD* task descriptions for specifying user inter-faces, at both semantic and presentation levels. In: Proceedings of Fourth Interna-tional Workshop on Design, Specification, and Verification of Interactive Systems, pp. 193–208 (1997)

24. Giese, M., Mistrzyk, T., Pfau, A., Szwillus, G., Detten, M.: Amboss: a task modeling approach for safety-critical systems. In: Forbrig, P., Paternò, F. (eds.) HCSE/TAMODIA-2008. LNCS, vol. 5247, pp. 98–109. Springer, Heidelberg (2008)
25. Green, M.: A survey of three dialogue models. Comput. Netw. **5**, 244–275 (1986)
26. Guerrero-Garcia, J., Gonzalez-Calleros, J.M., Vanderdonckt, J., Munoz-Arteaga, J.: A theoretical survey of user interface description languages: preliminary results. In: Proceedings of the 2009 Latin American Web Congress, LA-WEB 2009, pp. 36–43. IEEE Computer Society (2009)
27. Haklay, M., Singleton, A., Parker, C.: Web mapping 2.0: the neogeography of the geoweb. Comput. Netw. **2**(6), 2011–2039 (2008)
28. Harel, D., Marelly, R.: Come, Let's Play: Scenario-Based Programming Using LSC's and the Play-Engine. Springer-Verlag, New York (2003)
29. Hix, D., Hartson, H.R.: Developing User Interfaces: Ensuring Usability Through Product & Process. Wiley, Chichester (1993)
30. Huynh, D.F., Karger, D.R., Miller, R.C.: Exhibit: lightweight structured data publishing. In: Proceedings of the 16th International World Wide Web Conference, pp. 737–746 (2007)
31. Huynh, D.F., Mazzocchi, S., Karger, D.R.: Piggy bank: experience the semantic web inside your web browser. In: Proceedings of the International Semantic Web Conference (ISWC), pp. 413–430 (2005)
32. Koch, N., Knapp, A., Zhang, G., Baumeister, H.: UML-based web engineering. In: Rossi, G., Pastor, O., Schwabe, D., Olsina, L. (eds.) Web Engineering: Modelling and Implementing Web Applications. HCI, pp. 157–191. Springer, London (2008)
33. Kohler, H.-J., Nickel, U., Niereand, J., Zandorf, A.: Using UML as visual programming language. Technical report tr-ri-99-205 (1999)
34. Kolski, C.: Analyse et conception de l'IHM: Tome 1. Interaction homme-machine pour les SI, Hermès (2001)
35. Krasner, G.E., Pope, S.T.: A cookbook for using the model-view controller user interface paradigm in smalltalk-80. Comput. Netw. **1**, 26–49 (1988)
36. Limbourg, Q., Vanderdonckt, J., USIXML: a user interface description language supporting multiple levels of independence. In: ICWE Workshops, pp. 325–338 (2004)
37. Loustau, P., Nodenot, T., Gaio, M.: Design principles and first educational experiments of PIIR, a platform to infer geo-referenced itineraries from travel stories. Int. J. Interact. Technol. Smart Educ. **6**, 23–29 (2009)
38. Luong, T.N.: End-user centered modeling applied to interactive geographic application design: an approach based on contents and uses. Modélisation centrée utilisateur final appliquée à la conception d'applications interactives en géographie: une démarche basée sur les contenus et les usages. Ph.D. thesis, Université de Pau et des Pays de l'Adour (2012)
39. Luong, T.N., Etcheverry, P., Marquesuzaà, C.: An interaction model and a framework dedicated to web-based geographic applications. In: Proceedings of the International Conference on Management of Emergent Digital EcoSystems, MEDES 2011, pp. 235–242. ACM, New York (2011)
40. Luong, T.N., Etcheverry, P., Marquesuzaà, C., Nodenot, T.: A visual programming language for designing interactions embedded in web-based geographic applications. In: Proceedings of the 2012 ACM International Conference on Intelligent User Interfaces, IUI 2012, pp. 207–216. ACM (2012)

41. Luong, T.N., Etcheverry, P., Nodenot, T., Marquesuzaà, C.: WIND: an interaction lightweight programming model for geographical web applications. In: Bocher, E., Neteler, M. (eds.) Geospatial Free and Open Source Software in the 21st Century. Lecture Notes in Geoinformation and Cartography, pp. 211–225. Springer, Heidelberg (2009)

42. Luong, T.N., Etcheverry, P., Nodenot, T., Marquesuzaà, C., Lopistéguy, P.: End-user visual design of web-based interactive applications making use of geographical information: the WINDMash approach. In: Fifth European Conference on Technology Enhanced Learning, pp. 536–541 (2010)

43. Luong, T.N., Laborie, S., Nodenot, T.: A framework with tools for designing web-based geographic applications. In: Proceedings of the 11th ACM Symposium on Document Engineering, DocEng 2011, pp. 33–42. ACM (2011)

44. Luong, T.N., Marquesuzaà, C., Etcheverry, P., Nodenot, T., Laborie, S.: Facilitating the design, evaluation process of web-based geographic applications: a case study with WINDMash. In: Proceedings of Second International Conference on Future Data and Security Engineering, FDSE 2015, Ho Chi Minh City, Vietnam, 23–25 November 2015, pp. 259–271 (2015)

45. Mahfoudhi, A., Abed, M., Tabary, D.: From the formal specifications of user tasks to the automatic generation of the HCI specifications. In: Blandford, A., Vanderdonckt, J., Gray, P. (eds.) HCI 2001 and IHM 2001, pp. 331–347. Springer, London (2001)

46. Marion, C.: What is interaction design and what does it mean to information designers? (1999). http://mysite.verizon.net/resnx4g7/PCD/WhatIsInteractionDesign.html

47. Meliá, S., Gómez, J., Pérez, S., Díaz, O.: A model-driven development for GWT-based rich internet applications with OOH4RIA. In: Proceedings of the 2008 Eighth International Conference on Web Engineering, ICWE 2008, pp. 13–23. IEEE Computer Society, Washington, DC (2008)

48. Mori, G., Paterno, F., Santoro, C.: Design and development of multidevice user interfaces through multiple logical descriptions. IEEE Trans. Softw. Eng. **30**, 507–520 (2004)

49. Normand, V., Siroco, L.M.: de la spécification conceptuelle des interfaces utilisateur à leur réalisation. Ph.D. thesis, Thèse de doctorat Informatique préparée au Laboratoire de Génie Informatique (IMAG), Université Joseph Fourier 258 pages (1992)

50. Ormerod, T.C., Shepherd, A.: Using Task Analysis for Information Requirements Specification: The SGT Method, pp. 1–24. Lawrence Erlbaum Associates (2004)

51. Paillas, V.: Intérêt des applications "WIND" pour l'exploitation pédagogique de textes décrivant des itinéraires: les pratiques d'annotation au service du lire et interpréter différents langages. Master EFE - 2I2N - FEN de l'Université Toulouse 2 (IUFM Toulouse) (2011)

52. Pallota, V.: Computational dialogue models. Technical Report Report IM2.MDM-02, Faculty of Computer and Communication Sciences, Swiss Federal Institute of Technology - Lausanne (2003)

53. Paternò, F., Mancini, C., Meniconi, S.: Concurtasktrees: a diagrammatic notation for specifying task models. In: Proceedings of the IFIP TC13 Interantional Conference on Human-Computer Interaction, INTERACT 1997, pp. 362–369. Chapman & Hall Ltd., London (1997)

54. Pfaff, G.E. (ed.): User Interface Management Systems. Springer-Verlag, New York (1985)

55. Reiß, D., Rumpe, B.: Using lightweight activity diagrams for modeling and generation ofweb information systems. In: Mayr, H., Kop, C., Liddle, S., Ginige, A. (eds.) Information Systems: Methods, Models, and Applications. LNBIP, vol. 137, pp. 61–72. Springer, Heidelberg (2013)
56. Rogers, Y., Sharp, H., Preece, J.: Interaction Design: Beyond Human-Computer Interaction, 3rd edn. Wiley, Chichester (2011)
57. Sallaberry, C., Royer, A., Loustau, P., Gaio, M., Joliveau, T.: GeoStream: spatial information indexing within textual documents supported by a dynamically parameterized web service. In: Proceedings of the International Opensource Geospatial Research Symposium, OGRS 2009 (2009). http://hal.inria.fr/docs/00/45/19/49/PDF/ogrs.pdf
58. Samaan, K., Tarpin-Bernard, F.: Task models and interaction models in a multiple user interfaces generation process. In: Proceedings of the 3rd Annual Conference on Task Models and Diagrams, TAMODIA 2004, pp. 137–144. ACM, New York (2004)
59. Scapin, D., Pierret-Goldbreich, C.: Towards a method for task description: MAD. In: Berlinguet, L., Berthelette, D. (eds.) WWU 1989, pp. 27–34. Elsevier Science, North-Holland (1989)
60. Schwinger, W., Retschitzegger, W., Schauerhuber, A., Kappel, G., Wimmer, M., Prll, B., Cachero, C., Casteleyn, S., Troyer, O.D., Fraternali, P., Garrigs, I., Garzotto, F., Ginige, A., Houben, G.-J., Koch, N., Moreno, N., Pastor, O., Paolini, P., Pelechano, V., Rossi, G., Schwabe, D., Tisi, M., Vallecillo, A., van der Sluijs, K., Zhang, G.: A survey on web modeling approaches for ubiquitous web applications. IEEE Trans. Softw. Eng. **4**(3), 234–305 (2008)
61. Sears, A., Jacko, J.A. (eds.): The Human-Computer Interaction Handbook: Fundamentals, Evolving Technologies and Emerging Applications, 2nd edn., September 2007
62. Silva, P.P.D.: User interface declarative models and development environments: a survey. In: Palanque, P., Paternò, F. (eds.) DSV-IS 2000. LNCS, vol. 1946, pp. 207–226. Springer, Heidelberg (2001)
63. Tarby, J.-C., Barthet, M.-F.: The DIANE+ method. In: Vanderdonckt, J. (ed.) CADUI, pp. 95–120. Presses Universitaires de Namur (1996)
64. Veer, G.C.V.D., Lenting, B.F., Bergevoet, B.A.J.: GTA: groupware task analysis - modeling complexity. IEEE Trans. Softw. Eng. **91**, 297–322 (1996)
65. Vuillemot, R., Rumpler, B.: A web-based interface to design information visualization. In: Proceedings of the International Conference on Management of Emergent Digital EcoSystems, MEDES 2009, pp. 26:172–26:179. ACM (2009)
66. Wong, J., Hong, J.I.: Making mashups with marmite: towards end-user programming for the web. In: Proceedings of the SIGCHI Conference on Human Factors in Computing Systems (CHI), pp. 1435–1444 (2007)
67. Ziadi, T., Blanc, X., Raji, A.: From requirements to code revisited. In: Proceedings of the 2009 IEEE International Symposium on Object/Component/Service-Oriented Real-Time Distributed Computing, pp. 228–235. IEEE Computer Society (2009)

Applying Data Analytic Techniques
for Fault Detection

Ha Manh Tran[✉], Sinh Van Nguyen, Son Thanh Le, and Quy Tran Vu

Computer Science and Engineering, International University -
Vietnam National University, Ho Chi Minh City, Vietnam
{tmha,nvsinh,ltson,vtquy}@hcmiu.edu.vn

Abstract. Monitoring events in communication and computing systems becomes more and more challenging due to the increasing complexity and diversity of these systems. Several supporting tools have been created to assist system administrators in monitoring an enormous number of events daily. The main function of these tools is to filter as many as possible events and present highly suspected events to the administrators for fault analysis, detection and report. While these suspected events appear regularly on large and complex systems, such as cloud computing systems, analyzing them consumes much time and effort. In this study, we propose an approach for evaluating the severity level of events using a classification decision tree. The approach exploits existing fault datasets and features, such as bug reports and log events to construct a decision tree that can be used to classify the severity level of other events. The administrators refer to the result of classification to determine proper actions for the suspected events with a high severity level. We have implemented and experimented the approach for various bug report and log event datasets. The experimental results reveal that the accuracy of classifying severity levels by using the decision trees is above 80%, and some detailed analyses are also provided.

Keywords: Event monitoring · Fault data analysis · Fault detection · Classification decision tree · Software bug report

1 Introduction

The increasing complexity and diversity of communication and computing systems makes management operations more and more challenging. Cloud computing systems [1], as an example, facilitate computing resource management operations on large computing systems to provision infrastructures, platforms and software as services. Armbrust [2] has specified 10 hindrances for managing cloud systems and services. Several hindrances including service availability, performance unpredictability and failure control are closely involved with event monitoring, one of the main functions of fault management. Monitoring events on these systems usually deals with a large number of events. The system administrators needs the support of tools that filter out many events and keep non-trivial events. However, these systems provide so many non-trivial events that

© Springer-Verlag GmbH Germany 2017
A. Hameurlain et al. (Eds.): TLDKS XXXI, LNCS 10140, pp. 30–46, 2017.
DOI: 10.1007/978-3-662-54173-9_2

the administrators cannot handle. Furthermore, there is no guarantee that trivial events cannot cause system failure, e.g., warning events can become serious problems if there is no a proper action.

We have proposed an approach for evaluating the severity level of log events using classification and regression decision trees (CART trees). The idea of this approach is to determine the severity level of events automatically, thus providing the system administrators a decision whether further actions are needed for fault detection. The approach focuses on constructing a decision tree based on fault datasets and features, such as bug report and log events, and then using this tree to classify the severity level of other events. We have used bug report datasets obtained from existing bug tracking systems (BTSs) and log events obtained from monitoring systems to implement and experiment decision trees. The contribution is thus threefold:

1. Proposing an approach of using the classification decision tree for fault data analysis
2. Applying this approach to fault datasets for classifying the severity level of events
3. Providing the performance and efficiency evaluation of the approach on various fault datasets

In this paper, we have extended the previous study [3] to using large and real datasets of 500.000 bug reports and Mela log events for the evaluation of the proposed approach. The rest of the paper is structured as follows: the next section includes several analysis techniques applied to software maintenance, system failure and reliability, background of classification decision trees in data analysis. Section 3 describes the fundamentals of growing decision trees based on classification decision trees, focusing on entropy splitting rule and tree growing process. Some mathematical formulas and explanations are referred from the study of Breiman et al. [4]. Section 4 presents the characteristics of fault datasets and features. It also includes several processes of constructing decision trees for fault datasets. Several experiments in Sect. 5 report the performance and efficiency evaluation of the fault data analysis approach using the decision tree before the paper is concluded in Sect. 6.

2   Related Work

The authors of the study [5] have proposed an approach for analyzing fault cases in communication systems. The approach exploits the characteristics of semi-structured fault data by using multiple field-value and semantic vectors for fault representation and evaluation. Note that a fault case usually contains administrative field-value and problem description parts. The approach encounters the problem of high computation cost when processing semantic matrices for large fault datasets. Another study [6] from the same authors has reduced the computation problem by analyzing several types of fault classifications and relationships. This approach exploits package dependency, fault dependency, fault keywords, fault classifications to seek the relationships between fault causes. These approaches have been evaluated on software bug datasets obtained from different open source bug tracking systems. Sinnamon et al. [7] has applied the binary decision diagram to identify system failure and reliability. Large systems usually produce thousands of events that consume a large amount of processing time. This diagram associated with if-then-else rules and optimized techniques reduces time consuming problem. The study [8] has proposed an analysis strategy aiming at increasing the likelihood of obtaining a binary decision diagram for any given fault tree while ensuring the associated calculations as efficient as possible. The strategy contains 2 steps: simplifying the fault tree structure and obtaining the associated binary decision diagram. The study also includes quantitative analysis on the set of binary decision diagrams to obtain the probability of top events, the system unconditional failure intensity and the criticality of the basic events. The authors of the study [9] have presented two new tree-based techniques for refining the initial classification of software failures based on their causes. The first technique uses tree-like diagrams to represent the results of hierarchical cluster analysis. The second technique refines an initial failure classification that relies on generating a classification tree to recognize failed executions. This technique uses classification and regression tree for each subject of programs. Zheng et al. [10] has presented a decision tree learning approach based on the C4.5 algorithm to diagnose failures in large Internet sites. The approach records runtime properties of each request and applies automated machine learning and data mining techniques to identify the causes of failures. The approach has been evaluated on application log datasets obtained from the eBay centralized application logging framework. The study [11] proposes a nine-state model of adaptive behavior to enable fault detection in mobile applications. This model detects faults caused by erroneous adaptation process, and asynchronous update of context information, which leads to inconsistencies between the external physical context and internal representation within an application. The study [12] proposes a dynamic adaptation model that offers increased expressive power to compose complex adaptation rules, and guarantees soundness in fault detection. The recent study [13] introduces an elasticity analytic technique for cloud services. It also defines the concepts of elasticity space and elasticity pathway, and applies these concepts in evaluating the elasticity of cloud services. The Mela tool as the result of this study is an open source tool for monitoring and analyzing the elasticity of cloud services. This tool has been used in this study to collect log events.

Classification and regression trees (CART) [14] have been introduced by Breiman et al. and widely been used in data mining. Two main types of decision trees are classification and regression trees. The former tree predicts the outcome that belongs to one of the classes of the input data, e.g., predicting that today's weather is sunny, rainy or cloudy, while the later tree predicts the outcome that can be considered a real number, e.g., predicting that today's temperature is 25.3, 27.5, or 29.7 °C. Trees used for regression and classification have some similarities and also differences, such as the procedure used to determine where to split. There are several variants of decision tree algorithms. Iterative Dichotomiser 3 (ID3) [15] was developed in 1986 by J.R. Quinlan. This algorithm creates a multi-level tree that seeks a categorical feature for each node using a greedy method. The features yield the largest information gain for categorical targets. Trees are grown to their maximum size and then applied to generalize to unseen data. The algorithm C4.5 [16] is an extension of the ID3 algorithm that converts the trained trees as the output of the ID3 algorithm into sets of if-then rules. The accuracy of rules is evaluated by determining the order in which these rules are applied. This algorithm uses numerical variables to define a discrete attribute and partitions the continuous attribute values into a discrete set of intervals. It avoids finding categorical features. Chi-squared automatic inter-action detector (CHAID) [17] uses multi-level splits to compute classification trees. This algorithm focuses on categorical predictors and targets. It computes a chi-square test between the target variable and each available predictor and then uses the best predictor to partition the sample into segments. It repeats the process with each segment until no significant splits remain. There are several differences between the CHAID and CART algorithms: (i) CHAID uses the chi-square measure to identify splits, whereas CART uses the Gini or Entropy rule; (ii) CHAID supports multi-level splits for predictors with more than two levels, whereas CART supports binary splits only and identifies the best binary split for complex categorical or continuous predictors; (iii) CHAID does not prune the tree, whereas CART prunes the tree by testing it against an independent (validation) dataset or through n-fold cross-validation.

3   CART Approach

The CART approach [4] uses a binary recursive partitioning process to build a decision tree. This process starts with the root node where data features are split into two children nodes and each of the children node is in turn split into grandchildren nodes based on splitting rules. The process runs recursively until no further splits are possible due to lack of data features and the tree reaches a maximal size. The process deals with continuous and nominal features as targets and predictors.

3.1   Entropy Splitting Rule

A decision tree is built top-down from a root node and involves partitioning data into subsets that contain instances with similar values (homogeneous). The CART algorithm uses entropy to calculate the homogeneity of a sample.

$$H(S) = -\Sigma_{x \in X} P(x) log P(x) \tag{1}$$

where, S is the current (data) set for which entropy is being calculated. X is a set of classes in S. $P(x)$ is the proportion of the number of elements in class x to the number of elements in set S. When $H(S) = 0$ the set S is perfectly classified.

Information gain $IG(A, S)$ is the measure of the difference in entropy from before to after the set S is split on an attribute A. In other words, how much uncertainty in S was reduced after splitting set S on attribute A.

$$IG(A, S) = H(S) - \Sigma_{t \in T} P(t) H(t) \tag{2}$$

where, $H(S)$ is entropy of set S. T is the subset created from splitting set S by attribute A. $P(t)$ is the proportion of the number of elements in t to the number of elements in set S. $H(t)$ is entropy of subset t. Information gain can be calculated (instead of entropy) for each remaining attribute. The attribute with the largest information gain is used to split the set S on this iteration.

3.2   Tree Growing Process

The tree growing process uses a set of data features as input. A feature can be ordinal categorical, nominal categorical or continuous. The process chooses the best split among all the possible splits that consist of possible splits of each feature, resulting in two subsets of data features. Each split depends on the value of only one feature. The process starts with the root node of the tree and repeatedly runs three steps on each node to grow the tree, as shown in Fig. 1.

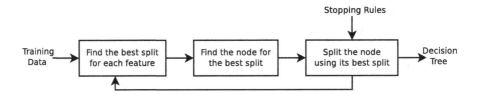

Fig. 1. A process of growing a CART decision tree

The first step is to find the best split of each feature. Since feature values can be computed and sorted to examine candidate splits, the best split maximizes the defined splitting criterion. The second step is to find the best split of the node among the best splits found in the first step. The best split also maximizes the defined splitting criterion. The third step is to split the node using its best split found in the second step if the stopping rules are not satisfied. Several stopping rules are used:

- If a node becomes pure; that is, all cases in a node have identical values of the dependent variable, the node will not be split.
- If all cases in a node have identical values for each predictor, the node will not be split.
- If the current tree depth reaches the user-specified maximum tree depth limit value, the tree growing process will stop.
- If the size of a node is less than the user-specified minimum node size value, the node will not be split.

– If the split of a node results in a child node whose node size is less than the user specified minimum child node size value, the node will not be split.

Figure 6 plots a sample CART tree with 4 levels (refer to the end of the paper). The tree grows enormously as the data size increases.

4 Fault Data Analysis

Fault data analysis in this study focuses on using a decision tree to evaluate the severity level of suspected fault cases, such as bug reports, log events or trace messages. We have used bug report datasets for analysis because bug reports are already verified while log events are usually not verified yet.

4.1 Bug Report Data

Bug report data contains software and hardware bug reports obtained from forums, archives and BTSs. Several tracker sites available on the Internet, such as Bugzilla [18], Launchpad [19], Mantis [20], Debian [21] provide web interfaces to their bug data. Tracker sites differ from data inclusion and presentation, but share several similar administration and description fields. While the administration fields are represented as field-value pairs, such as severity, status, platform, content, component and keyword, the problem description field details the problem and follow-up discussions represented as textual attachments. We have used a web crawler to get as much access to bug data as ordinary users. The crawler retrieves the HTML pages of bug reports, then few parsers extract the content of bug reports and save the content to a database following a unified bug schema [22]. Table 1 reports popular BTSs and numbers of downloadable bug reports for tracker sites.

A bug report contains several features shown in the unified bug schema [22]. Some features cause less impact on determining the severity of the bug report,

Table 1. Popular bug tracking sites (as of November 2014). A plus indicates that the numbers present a lower bound

Tracker site	System	Bugs
bugs.debian.org	Debian BTS	900.000$^+$
bugs.kde.org	Bugzilla	400.000$^+$
bugs.eclipse.org	Bugzilla	400.000$^+$
bugs.gentoo.org	Bugzilla	350.000$^+$
bugzilla.mozilla.org	Bugzilla	800.000$^+$
bugzilla.redhat.com	Bugzilla	900.000$^+$
qa.netbeans.org	Bugzilla	250.000$^+$
bugs.launchpad.net	Launchpad	1.200.000$^+$

Table 2. List of important features

Feature	Description	Data types
Status	The open, fixed or closed status of the bug	Enumerate
Component	The component contains the bug	Enumerate
Software	The software contains the bug	Enumerate
Platform	The platform where the bug occurs	Enumerate
Keyword	The list of keywords that describe the bug	Text
Relation	The list of bugs related to the bug	Numeric
Category	The category of the bug	Enumerate

such as owner, created time, updated time, etc. Our approach therefore focuses on the features as shown in Table 2. Note that each bug report contains the severity feature with a value. It is necessary to ignore this feature when building the tree to avoid some side effect. The keyword feature that contains the description and discussion of the bug requires further data processing.

4.2   Data Processing

Processing features improves the quality of the training datasets and thus enhance the performance of the decision tree. A bug report contains a textual part of the problem description and some discussions that hide distinct keywords or groups of keywords. We have applied the term frequency–inverse document frequency (tf×idf) method to reveal these keywords for the keyword feature. This method measures the significance of keywords to bug reports in a bug dataset by the occurrence frequency of the keywords in a bug report over the total number of the keywords of the bug report (term frequency) and the occurrence frequency of the keywords in other bug reports over the total number of bug reports (inverse document frequency). A distinct group of keywords contains related keywords with high significance. As a consequence, the keyword feature includes a set of keywords and groups that best describe the bug report. However, since bug reports are obtained from various BTSs, their descriptions and discussions contain redundant words, nonsense words or even meaningless words, such as: memory address, debug information, system path, article, etc. Algorithm 1 filters out these words from the bug dataset. We have implemented this algorithm in Python programming language.

The first step is to load the bug dataset focusing on the keyword feature. The next three steps are to filter useless keywords. The stop-word set is the set of popular keywords that usually appear in textual description such as a, an, the, of, etc. The regular expression contains characters [0–9], [a-f] and [A-F], while the special characters contains _, -, \. The final step is to apply the tf×idf method on the whole keyword set and remove trivial keywords, i.e., keywords with low tf×idf values.

Algorithm 1. Filtering keywords for a bug dataset

Input   : Raw keyword set
Output: Filtered keyword set
1 Load raw keyword set;
2 Remove duplicated words and redundant words by using stop-word set;
3 Remove meaningless words by using regular expression;
4 Remove memory addresses by filtering special characters;
5 Process tf×idf on the whole keyword set;
6 **return** Filtered keyword set;

4.3   Tree Construction

The previous section explains using Entropy splitting rule to grow a decision tree. We present in this section using Scikit Learn library [23] to construct decision trees for bug datasets. Scikit Learn is an open source machine learning library for Python programming language and provides several classification, regression and clustering algorithms. It is designed to interoperate with Python numerical and scientific libraries such as NumPy [24] and SciPy [25]. The CART algorithm is one of the main classification algorithms supported by Scikit Learn. Algorithm 2 presents main steps to construct decision trees using the Scikit Learn library:

Algorithm 2. Constructing a decision tree for a bug dataset

Input   : Processed bug dataset
Output: Decision tree
1 Load the dataset into pandas data-frame and drop the platform feature;
2 Factorize the features;
3 Load sample data and class label;
4 Split the dataset into the training set and testing set;
5 Fit the training set into decision tree classifier;
6 Construct the tree using entropy criterion;

The first step is to load the dataset into pandas data-frame that is a special tabular data structure to prepare data for the CART algorithm. It is also important to drop the platform feature in the data-frame because the dataset is already grouped by this feature. Since the CART algorithm cannot deal with non-numerical values, while the feature values in the bug dataset are non-numeric, i.e., enumerate or text, all the feature values need to be factorized into numerical values in the second step. The pandas library supports for converting non-numerical values to numerical values. Each distinct value is replaced by a unique integer, e.g., the severity feature contains 4 values: feature, minor, normal and critical corresponding to 0, 1, 2, 3 after factorization. The next step is to separate the data-frame into 2 parts. The first part is the sample data that contains the numerical values of all features, while the second part is the class label that marks the numerical classes for each particular bug. The most important step in this algorithm is to partition the sample data and class label into the training set and testing set. The training set is used for training the decision tree, while the testing set is used for evaluating the decision tree. The percentages of the training and testing sets are 75% and 25% respectively. Finally, the decision tree is trained by a method supported by Scikit Learn library. The input of this method is the training set found in the previous step. Figure 7 plots a part of a decision tree for the Linux platform dataset (refer to the end of the paper).

Since the decision tree contains multiple levels, we only present the first 4 levels. The leaf nodes contains the following values:

1. The first component counts samples that have the severity of feature
2. The second component counts samples that have the severity of critical
3. The third component counts samples that have the severity of minor
4. The fourth component counts samples that have the severity of normal

5   Evaluation

We have used a dataset of 500.000 bug reports approximately for experiments. A large dataset usually yields a large decision tree that possibly causes the performance problem due to the complexity and memory consumption of the tree. The authors of the study [26] have already proposed an approach to construct decision trees from large datasets. This approach builds a set of decision trees based on tractable size training datasets which are the subsets of the original dataset. As a result, the approach reveals the over-fitting problem of large decision trees. We have separated bug reports into 4 smaller datasets following the platform feature: 180.000 bug reports occurring on all platforms (All platform), 160.000 bug reports occurring on Windows platform (Win platform), 80.000 bug report occurring on Linux platform (Linux platform) and 80.000 bug report occurring on Macintosh platform (Mac platform). We have built several decision trees and performed all experiments on laboratory workstations with Intel Core i5-4590 CPU 3.30 GHZ, 8 GB of RAM and Ubuntu 14.04 LTS. In addition, we have used the Mela tool to collect log events with an interval of 5 s. The interval is sufficient to capture changes on the monitored system, but it also produces several duplications on idle time. We have filtered duplicated events in the Mela dataset for experiments.

The first experiment measures time consumption for constructing decision trees over various datasets. Time consumption linearly increases as the size of datasets increases, as shown in Fig. 2. It takes 340 ms to 380 ms approximately to build a decision tree of 160.000 bug reports. Note that time consumption depends on numbers of events and features of bug reports. Bug reports in the Win platform dataset contain less one feature than bug reports in the whole dataset, i.e., the platform feature is eliminated in a platform specific dataset, thus time consumption for both datasets is slightly different. It takes 900 ms approximately to construct a decision tree of 500.000 bug reports. However, log files usually contain millions of events, reducing processing time is an important task.

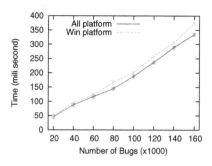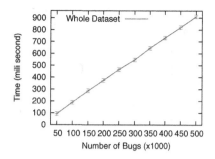

Fig. 2. Time consumption for constructing decision trees over various datasets

The datasets contain thousands of bug reports that possibly miss several features. It is necessary to apply the median imputation method for these datasets to fill the missing features. Dealing with the missing values is one of the most common problems in data training process. This problem occurs when data values are unavailable for observations due to the lack of responses: data is provided for neither several features nor a whole case. Lacking values are sometimes caused by researchers due to collecting data improperly or making mistakes in data input.

Using training datasets with the missing values can affect accuracy in classification. Several prevailing methods that are capable of dealing with this issue have been developed before. Case deletion method discards cases with the missing values for at least one feature. A variant of this method only eliminates cases with a high level of the missing values while determining the extent of features for cases with a low level of the missing values. Mean imputation method replaces the missing values for features by the mean of all known values of these features in the class to which the case with the missing values belongs. Similar to the mean imputation method, the median imputation method replaces the missing values for features by the median of all known values of the features in the class to which the case with missing values belongs. Using median avoids the presence of outliers and also assures the robustness of the method. This method is suitable for datasets that the distribution of the values of a certain feature is skewed. Modified K-nearest neighbor method determines the missing values for a case by considering a certain number of the most similar cases. The similarity of two cases is measured by a distance function.

The second experiment fills the missing values for bug reports using the mean imputation method. It then compares cross-validation scores for both datasets with and without imputation. Figure 3 on the left side reports the difference of cross-validation scores for the All platform dataset with and without imputation. The All platform dataset with imputation obtains the average cross-validation score of 0.68 that improves considerably a number of the missing values from the All platform dataset without imputation. Especially with imputation, the All platform dataset of 160.000 bug reports reaches the cross-validation score of 0.7.

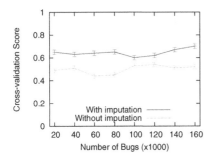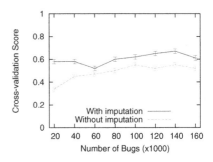

Fig. 3. Cross-validation comparison for the All platform (left) for the Win platform (right) with and without imputation

Similarly, the Win platform dataset with imputation obtains the average cross-validation score of 0.62 that slightly improves a number of the missing values from the Win platform dataset without imputation, as shown in Fig. 3 on the right side. With imputation, the Win platform dataset of 160.000 bug reports reaches the cross-validation score of 0.6. The Win platform dataset performs worse than the All platform dataset. We observe that the missing values of bug reports in the All platform dataset are less specific than that the missing values of bug reports in the Win platform dataset. Note that the number of the missing values increases as the size of datasets increases, thus using the imputation technique can improve the accuracy score of classification.

The third experiment evaluates the accuracy of the decision trees using various sizes of datasets. The idea is to divide the original dataset with imputation into the training and testing datasets. While the training dataset is used to build a decision tree, the testing dataset is used to evaluate the accuracy of this decision tree. The extreme case of cross-validation, namely leave-one-out cross-validation, has been used for this experiment. We have used the decision trees to classify bug reports from the testing dataset into severity levels, then compared these classified severity levels with the correct severity levels of the testing dataset. Accuracy score is calculated based on the number of matching severity levels.

Figure 4 on the left side reports high and stable accuracy scores for the All and Win platform datasets with the average score of 0.9 approximately. Both datasets perform similarly. We observed that bug reports for these platforms tend to be common problems that can be easily reproduced, determining severity levels for these bug reports is rather straightforward and precise. However, some bug reports from a specific platform are very specific and difficult to be classified into severity levels properly. These bug reports reduces accuracy scores. Figure 4 on the right side also presents the similar accuracy scores of the Linux and Mac platform datasets with the average score of 0.85 approximately. The All and Win platform datasets are larger than the Mac and Linux platform datasets that possibly cause an impact on accuracy scores as the size of these datasets

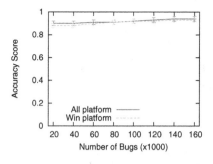
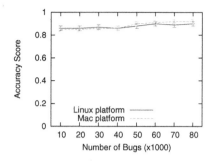

Fig. 4. Accuracy comparison between the All and Win platforms (left) and between the Linux and Mac platforms (right)

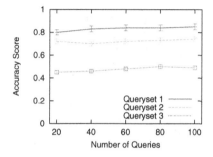

Fig. 5. Accuracy of the decision tree constructed by the whole dataset using various querysets

increases. In addition, average time to train a decision tree of 160.000 bug reports is considerably fast.

The last experiment measures the accuracy of the decision tree constructed by the whole dataset using three querysets. The first queryset contains queries extracted from the testing dataset. The second queryset contains queries extracted from the existing BTSs. These queries are bug reports in open status and some features, such as, keyword, relation, category and severity are insufficient. The last queryset contains queries extracted from some monitoring tools, such as Nagios [27], Ganglia [28] and Mela [13]. These queries are only warning and error events that possess different features from bug reports. Figure 5 presents the accuracy score of the decision tree with three querysets. The first queryset outperforms the other querysets and its accuracy score considerably matches with the results of the above experiments because queries are from the testing dataset. The accuracy score decreases as the decision tree is large. The second queryset obtains the average accuracy score of 0.7. Queries from this queryset contain the same bug format with incomplete features, but still achieve reasonable accuracy scores. The third queryset performs poorly with the average accuracy score lower than 0.5. There are several differences in features between bug reports and log events that prevent the decision tree from

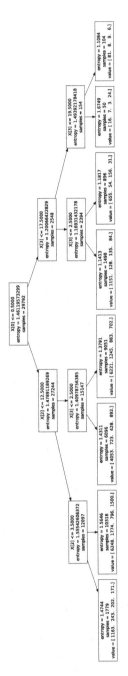

Fig. 6. A sample CART tree

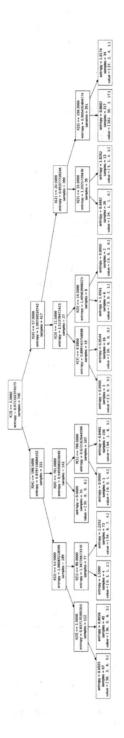

Fig. 7. Decision tree for the Linux platform dataset

classifying these queries precisely. Log events need additional data referred to as context-ware data including system load, network status, memory usage, number of processes, etc. to fulfill the missing features.

6   Conclusions

We have proposed an approach of using the CART decision tree for fault data analysis that can be applied to monitoring and detecting faults in communication networks and distributed systems. The log event data is so huge that system administrators and even supporting tools might miss critical signs, messages or events accidentally. This decision tree is characterized by the capability of learning from the training fault dataset and then determining the severity level of log events from the testing dataset. We have used a bug report dataset for evaluating the decision tree. Bug reports obtained from BTSs are to some extent related to log events with a severity level. Evaluating the approach focuses on the performance and efficiency of the decision tree. We have computed the time consumption of constructing the decision tree, the accuracy of classification and the imputation of the missing features. The experimental results reveal that the accuracy of classifying severity levels by using the decision trees is above 80%, especially 90% for the All platform dataset. Applying methods to deal with the missing features in the training dataset improves efficiency. Moreover, the decision tree with a tractable size training dataset consumes less processing time and possibly yields high efficiency. The decision tree constructed by bug reports does not perform well with log events due to the lack of significant features. Future work focuses on exploiting more common features in bug reports or log events, especially exploiting distinct keywords from textual description. While monitoring systems, log events can be extended to include additional context-aware features, such as system load, network status, memory usage, number of processes, etc. to evaluate precisely the severity level of log events [12,13].

Acknowledgements. This research activity is funded by Vietnam National University in Ho Chi Minh City (VNU-HCM) under the grant number B2017-28-01 (the type-B project "Augmenting fault detection services on large and complex network systems using context-aware data analysis")

References

1. Buyya, R., Yeo, C.S., Venugopal, S., Broberg, J., Brandic, I.: Cloud computing and emerging IT platforms: vision, hype, and reality for delivering computing as the 5th utility. Future Gener. Comput. Syst. **25**(6), 599–616 (2009)
2. Armbrust, M., Fox, A., Griffith, R., Joseph, A.D., Katz, R., Konwinski, A., Lee, G., Patterson, D., Rabkin, A., Stoica, I., Zaharia, M.: A view of cloud computing. ACM Commun. **53**(4), 50–58 (2010)
3. Tran, H.M., Nguyen, S., Le, S.T., Vu, Q.T.: Fault data analytics using decision tree for fault detection. In: Dang, T.K., Wagner, R., Küng, J., Thoai, N., Takizawa, M., Neuhold, E. (eds.) FDSE 2015. LNCS, vol. 9446, pp. 57–71. Springer, Heidelberg (2015). doi:10.1007/978-3-319-26135-5_5

4. Breiman, L., Friedman, J.H., Olshen, R.A., Stone, C.J.: Classification and Regression Trees. Chapman & Hall/CRC, New York (1984)
5. Tran, H.M., Schönwälder, J.: Fault representation in case-based reasoning. In: Clemm, A., Granville, L.Z., Stadler, R. (eds.) DSOM 2007. LNCS, vol. 4785, pp. 50–61. Springer, Heidelberg (2007). doi:10.1007/978-3-540-75694-1_5
6. Tran, H.M., Le, S.T., Ha, S.V.U., Huynh, T.K.: Software bug ontology supporting bug search on peer-to-peer networks. In: Proceedings of 6th International KES Conference on Agents and Multi-agent Systems Technologies and Applications (AMSTA 2013). IOS Press (2013)
7. Sinnamon, R.M., Andrews, J.D.: Fault tree analysis and binary decision diagrams. In: Proceedings in Reliability and Maintainability Annual Symposium, pp. 215–222 (1996)
8. Reay, K.A., Andrews, J.D.: A fault tree analysis strategy using binary decision diagrams. Reliab. Eng. Syst. Saf. **78**(1), 45–56 (2002)
9. Francis, P., Leon, D., Minch, M., Podgurski, A.: Tree-based methods for classifying software failures. In: Proceedings of 15th International Symposium on Software Reliability Engineering (ISSRE 2004), pp. 451–462, Washington, DC, USA. IEEE (2004)
10. Zheng, A.X., Lloyd, J., Brewer, E.: Failure diagnosis using decision trees. In: Proceedings of 1st International Conference on Autonomic Computing (ICAC 2004), Washington, DC, USA, pp. 36–43. IEEE Computer Society (2004)
11. Sama, M., Rosenblum, D.S., Wang, Z., Elbaum, S.: Model-based fault detection in context-aware adaptive applications. In: Proceedings of 16th ACM SIGSOFT International Symposium on Foundations of Software Engineering, New York, NY, USA, pp. 261–271. ACM (2008)
12. Xu, C., Cheung, S.C., Ma, X., Cao, C., Jian, L.: Detecting faults in context-aware adaptation. Int. J. Softw. Inf. **7**(1), 85–111 (2013)
13. Moldovan, D., Copil, G., Truong, H.L., Dustdar, S.: MELA: monitoring and analyzing elasticity of cloud services. In: Proceedings of 5th International Conference on Cloud Computing, pp. 80–87. IEEE Press (2013)
14. Breiman, L., Friedman, J., Stone, C., Olshen, R.: Classification and Regression Trees. Chapman & Hall, New York (1984)
15. Quinlan, J.R.: Induction of decision trees. Mach. Learn. **1**(1), 81–106 (1986)
16. Quinlan, J.R.: C4.5: Programs for Machine Learning. Morgan Kaufmann Publishers, San Francisco (1993)
17. Kass, G.V.: An exploratory technique for investigating large quantities of categorical data. Appl. Stat. **29**(2), 119–127 (1980)
18. Mozilla bug tracking system. https://bugzilla.mozilla.org/. Accessed Jan 2015
19. Launchpad bugs. https://bugs.launchpad.net/. Accessed Jan 2015
20. Mantis bug tracker. https://www.mantisbt.org/. Accessed Jan 2015
21. Debian bug tracking system. https://www.debian.org/Bugs/. Accessed Jan 2015
22. Tran, H.M., Lange, C., Chulkov, G., Schönwälder, J., Kohlhase, M.: Applying semantic techniques to search and analyze bug tracking data. J. Netw. Syst. Manag. **17**(3), 285–308 (2009)
23. Pedregosa, F., Varoquaux, G., Gramfort, A., Michel, V., Thirion, B., Grisel, O., Blondel, M., Prettenhofer, P., Weiss, R., Dubourg, V., Vanderplas, J., Passos, A., Cournapeau, D., Brucher, M., Perrot, M., Duchesnay, E.: Scikit-learn: machine learning in python. J. Mach. Learn. Res. **12**, 2825–2830 (2011)
24. Oliphant, T.: A Guide to NumPy, vol. 1. Trelgol Publishing, Spanish Fork (2006)
25. Silva, F.B.: Learning SciPy for Numerical and Scientific Computing. Packt Publishing, Birmingham (2013)

26. Hall, L.O., Chawla, N., Bowyer, K.W.: Decision tree learning on very large data sets. In: Proceedings of IEEE International Conference on Systems, Man and Cybernetics, vol. 3, pp. 2579–2584. IEEE (1998)

27. The industry standard in IT infrastructure monitoring (1999). http://www.nagios.org/. Accessed Nov 2015

28. Ganglia monitoring system (2000). http://ganglia.info/. Accessed Nov 2015

Protecting Biometrics Using Fuzzy Extractor and Non-invertible Transformation Methods in Kerberos Authentication Protocol

Thi Ai Thao Nguyen[✉] and Tran Khanh Dang

Faculty of Computer Science and Engineering, Ho Chi Minh City University of Technology,
VNU-HCM, Ho Chi Minh City, Vietnam
{thaonguyen,khanh}@hcmut.edu.vn

Abstract. Kerberos is a distributed authentication protocol which guarantees the mutual authentication between client and server over an insecure network. After the identification, all the subsequent communications are encrypted by session keys to ensure privacy and data integrity. Nowadays, many traditional authentication systems have tried moved to biometric system for convenience. However, the security and privacy of these systems need to put on the table. In this paper, we have proposed an efficient hybrid approach for protecting biometrics in remote authentication protocol based on Kerberos scheme. This protocol is not only resistant against attacks on the insecure network such as man-in-the-middle attack, replay attack,... but also able to protect the biometrics for using fuzzy extractor and non-invertible transformation. These techniques conceal the user's cancelable biometrics into the cryptographic key called biometric key. This key is used to verify a user in authentication phase. Therefore, there is no need to store users' plaint biometrics in the database. Even if biometric key is revealed, it is impossible for an attack to infer the users' biometrics for the high security of the fuzzy extractor scheme. Moreover, another remarkable contribution of this work is that a user can also change his biometric key without replacing his biometrics. The protocol supports multi-factor authentication to enhance security of the entire system.

Keywords: Biometrics · Remote authentication · Kerberos · Fuzzy extractor · Cancelable transformation · Mutual authentication

1 Introduction

In recent years, electronic commerce and especially mobile commerce have an exceeding growth. More and more services become online for their convenience. People, nowadays, can stay home, hold their mobile phone to do a lot of things such as: shopping, paying their bills, learning, meeting, chatting with friends, voting, checking in for millions of real life services like airports, hotels, schools, hospitals,... However, benefits always come along with challenges. The urgent issue for the mobile commerce is security. The user needs to be sure that their essential information is not revealed, or tracked. Therefore, the first step before using any m-commerce services is to verify the

© Springer-Verlag GmbH Germany 2017
A. Hameurlain et al. (Eds.): TLDKS XXXI, LNCS 10140, pp. 47–66, 2017.
DOI: 10.1007/978-3-662-54173-9_3

valid server. To server side, it obviously has to verify the users who request its service. And that is how our story about authentication begins.

Username/password is the traditional authentication method which is widely used by most m-commerce services. Nevertheless, this method exists some natural setbacks. It is obvious that a strong password is difficult for people to remember but easy for computer to guess. Especially, with recent technology development, the computer ability is being enhanced; meaning password cracking chance is rising too. Moreover, nothing guarantees that the person having the password is the right person. For that reason, biometric based authentication method was born; with its advantages, this method is gradually replacing its predecessor. The first advantage to be mentioned is that biometrics (such as face, voice, iris, fingerprint, palm-print, gait, signature,...) reflects a specific individual which helps preventing multi-user usage from one account [1]. This means a lot in case of lost mobile devices. There are hundreds of thousands of portable mobile devices are lost every year. And the problem is that these devices are not able to distinguish between the attackers who have the password with the genuine users. It is very dangerous because mobile devices store lots of private information of users. The next advantage is that using biometric method is more convenient for users because they do not have to remember or bring it with them all day long. For these reasons, there are more and more users choosing biometrics instead of password. In fact, many technologies have been released to support biometric based authentication system.

However, biometrics still has some troubles with the security of the authentication system. As we all know, human has a limited number of biometric traits; therefore, we cannot change our biometrics frequently like password once we suspect that the templates are revealed [2]. Moreover, the fact that people register many online services makes them use the same biometrics to sign in some services. That leads to the cross-matching attacks when attackers follow user's biometric template cross the online services in order to track their activities. Another concern relates to the natural set-backs of biometrics. The fact that biometrics reflects a specific individual means it contains sensitive information which users do not want attacker or even the server storing users' authentication data to discover. Last but not least, the network security needs to be discussed when user's private information is transmitted over insecure network [3].

The goal of this work is to propose the biometric-based remote authentication protocol relied on Kerberos - the secure distributed authentication protocol. This protocol not also guarantees the scalability property but also protects the sensitive data of users from attackers, and allows users to change authentication data over and over like password without replacing their biometrics. This is the remarkable contribution in comparison with the work in the conference paper [4].

The remaining parts of this paper are organized as follows. In the Sect. 2, related works is briefly reviewed. We show what previous works have done and their limitations. From that point, we present our motivation to fill the gap. In Sect. 3, we introduce the preliminaries used in the proposal. In the next section, our proposed protocol is described in detail. In the Sect. 5, the security analysis is presented to demonstrate for our proposal. Finally, the conclusion is included in the Sect. 6.

2   Related Works

Along with incredible growth of technology, many kinds of biometrics are applied in authentication systems. However, in mobile computing environment which supports the camera and recorder, the authentication system for m-commerce service often employs face, fingerprint, voice, or sometimes iris. For example, the authors in [5] presented a fingerprint based bio-cryptographic security protocol designed for client/server authentication. Fingerprint also applied in BioPKI authentication system [6]. In [7, 8], face was used as authentication factor in secure online service. Voice also was applied in [9]. In some protocols, they combined many kinds of biometrics in authentication system to enhance the accuracy [1, 10–13]. All these works, the authors tried to create a biometric based remote authentication protocols which supplied their users with convenience and privacy.

The concerns of security and privacy in remote architecture also attract the interest of many scientists [2, 14–19]. The first mentioned here is how to protect biometric template against the curiosity of the outside attacker or even the administrators of the remote servers. The detailed survey of various biometric template protection schemes was presented in [17] by Jain et al. This work also discussed the strengths and weaknesses of all techniques. Over the years, there have been plenty of techniques proposed to protect many kinds of biometric traits. Each of them is suitable for a certain biometric traits. However, they can be classified into two approaches, including feature transformation and biometric cryptosystem (as illustrated in Fig. 1). The first approach identified as feature transform allows users to replace a compromised biometric template while reducing the amount of information revealed. However, some methods of the approach cannot achieve an acceptable performance; others are unrealistic under assumptions from a practical view point [3]. The remarkable technique belongings this approach is non-invertible transformation. By employing one-way functions, a biometrics can be transformed into a secure version which is still fit for authentication purpose. For example, non-invertible transformation for fingerprint is proposed in [20]; face and voice others which can be extracted into vector form in [16, 21];... The second approach is biometric cryptosystem. It tries to combine the biometrics and cryptography technique in order to take advantages of both. These techniques employing these methods aim at generating a key, which derived from the biometric template or bound with the biometric template, and some helper data. Neither biometrics nor key is stored in client side. Only the helper data is needed to reproduce the original biometrics or the secret key. However, the biometric cryptosystem approach falls into the natural setback of biometrics. It lacks of cancelability property which allows users to revoke a compromised template and reissue a new one from the same biometrics. For this reason, some researches have proposed hybrid schemes which combine biometric cryptosystem with non-invertible transformation to take advantages of both approaches. The combination of secure sketch and ANN (Artificial Neural Network) was proposed in [15]. The fuzzy Vault was combined with Periodic Function-Based Transformation in [14], or with the non-invertible transformation to conduct a secure online authentication in [7]. The homomorphic cryptosystem was employed in fuzzy commitment scheme to achieve the blind authentication in [22]. The fuzzy commitment and the non-invertible transformation were integrated in [21]. As a result, we present the combination of fuzzy extractor and non-invertible transformation to protect biometrics in remote authentication protocol.

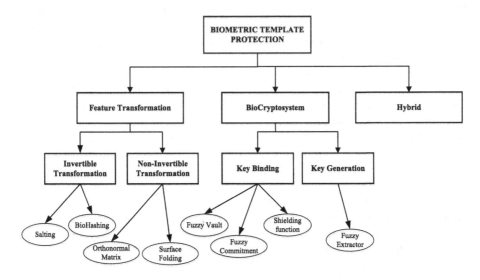

Fig. 1. Biometric template protection techniques.

To protect user's privacy through insecure network and the curiosity of a server, many biometric-based remote authentication protocols tried to obtain the mutual authentication requirement where both user and server can ensure they are communicating with the right side. In [8], Hisham et al. presented another approach that combined steganography and biometric cryptosystem in order to obtain the secure mutual authentication and key exchange between client and server in remote architecture. However, the authors did not pay much attention to the inside attacks. Therefore, server can take advantage of the user's authentication data to impersonate its clients and do some illegal transactions.

In [23], Lee-Hsu's protocol was proposed. The remarkable contribution of this work was the authors employed the idea of Chebyshev polynomial to obtain the mutual authentication between server and its clients. In 2015, Zhang Min et al. pointed out some design flaws in Lee-Hsu's protocol and proposed an enhanced scheme based on fuzzy extractors in [24]. Nonetheless, these protocols did not mention the resistance to the inside attacks. To deal with this problem, there have been few works published.

In 2010, blind authentication was proposed [3]. This protocol applied homomorphic cryptography to protect user's privacy against the violation of a server itself. Though the server could verify their users, it was not able to identify their users. It meant that when a user sent a login request, the server only could define he is a legal user or not without knowing who he is, what his purpose is. The homomorphic cryptography was also employed in eSketch protocol [22]. These kinds of protocols seemed to reach the perfect security. However, the price paid to achieve that level of security is unaffordable. The high computational complexity and longtime consumption caused these protocols impracticable. In addition to that, when the number of users increased, the time consumption was even longer. The number of users was also mentioned in the scalability property which has not been considered much yet in the previous works. When the number of users and the number of servers is growing, the number of templates which belongs to a user is large, and each server has to remember every user's template. That design makes the system vulnerable and wastes our resources.

To guarantee the scalability properties, Fengling et al. presented a biometric based remote authentication which employed the Kerberos protocol [19]. A biometric-Kerberos authentication protocol was suitable for e-commerce applications. The benefit of Kerberos is that expensive session-based user authentication can be separated from cheaper ticket-based resource access. However, the Achilles' heel of the proposed protocol is the biometric template was not protected. If an attacker penetrates into the KDC, he can extract the users' biometrics.

In this work, we propose the biometric-based remote authentication protocol emulated the Kerberos protocol. The contribution is the embedment fuzzy extractor and non-invertible transformation on Kerberos to obtain the security and the cancelability requirements of a biometric based authentication protocol. This protocol not also guarantees the scalability property but also protects the private information of users from attackers. The mutual authentication and the key agreement are also guaranteed in this protocol.

3   Preliminaries

3.1   Kerberos Protocol

Kerberos is a distributed authentication protocol which allows a process to excute on behalf of a user in order to authenticate the identity of a user to a verifier (authentication server) without sending any sensitive information of user over the insecure network. Kerberos provides integrity and security for the data which is sent between a client and a server. The protocol also immunises against eavesdropping and replay attacks. The design goals of this protocol aims at a client – server model and mutual authentication, both client and server are able to verify each other's identity.

In particular, Kerberos is not effective against password guessing attack. If a user creates a simple password, an attacker can totally figure out and use it to impersonate user. Moreover, Kerberos requires a trusted path for passwords to be transfered. If a user gives his or her password to a program which has been modified by attackers, or the path between a user and initial authentication server is monitored, it is easy for an attacker to obtain sufficient information and access the system on behalf of the user illegally. That is the reason why this protocol needs to be combined with other techniques for example: encryption, timestamp,... to address these limitations.

The Kerberos authentication protocol employs a series of encrypted messages to prove to a verifier that a client is running on behalf of a particular user. It is partly based on the Needham and Schroeder authentication protocol [25] with some modifications. These changes are the use of time-stamp to protect exchanged messages between client and server from replay-attack, and the addition of a ticket-granting service to support subsequent authentication without re-entry of a user's authentication information (for example password).

The history of Kerberos development is mentioned in detail in [26]. In this section, we represent how it works. The Kerberos authentication protocol has three entities: client, KDC – Key Distributed Center, and RS – Resource Service. Client is the user who has registered before. KDC includes database which stores user's data, AS-Authentication Server which is responsible for authenticate user and provides necessary information for user to verify this server, and the last one, TGS – Ticket Granting Server which is in charge of giving a ticket for a verified user. A user can access to Resource Service if and only if he or she possesses this ticket. The original Kerberos has three main phases (Fig. 2).

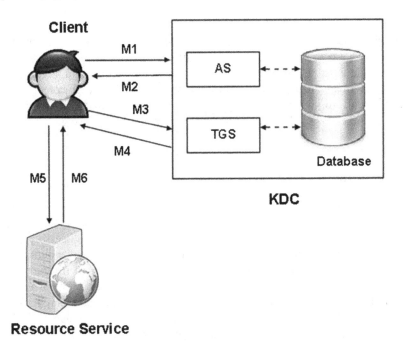

Fig. 2. The original Kerberos authentication protocol.

Phase 1. Client verifies AS.

Client sends a request M_1 to AS. M_1 contains userID, and timestamp. Then, AS checks whether the userID was stored in the database or not. If yes, AS create a secret key K_U from the user's password and use it to encrypt the session-key K_S which has just been created by AS. After that, AS sends a message M_2 to client. M_2 contains two messages A and B. A is the encryption of the session-key K_S using the secret key K_U. B is a granting ticket which is encrypted by TGS key K_{TGS}.

$$M_2 = \{A, B\}$$

$$A = Enc(K_s, K_U)$$

$$B = Enc((userID, timestamp, K_s), K_{TGS})$$

Once receiving M$_2$, client calculates the secret key K_U from his or her password in order to decrypt A and obtains the session key K_S.

Phase 2. TGS verifies Client. If successful, TGS grants ticket to client.

Client sends message M3 to TGS. M2 contains service request C and client authentication message D.

$$M_3 = \{C, D\}$$

$$C = (B, service\_ID)$$

$$D = Enc((userID, timestamp), K_s)$$

TGS uses its secret key K_{TGS} to decrypt B and obtains the session key K_S, then uses K_S to decrypt D and finds userID and timestamp. Finally, TGS compares userID and timestamp from messages B and D to verify the client. If they match, the client is authenticated, then TGS sends M4 to client. M4 contains the ticket E for client to access the resource service, which is encrypted by the service key K_{AP}, and the session key K_{C-AP} between the client and the resource service, which is also encrypted by K_S

$$M_4 = \{E, F\}$$

$$E = Enc((userID, timestamp, K_{C-AP}), K_{AP})$$

$$F = Enc(K_{C-AP}, K_s)$$

Once receiving M4, the client uses K_S to decrypt F and gets the session key K_{C-AP} shared with the resource service.

Phase 3. RS verifies client's ticket, and then establishes the communication session with client.

The client sends M5 to the RS. M5 contains the ticket E from the phase 2, and a message G. After receiving M5, RS employs its secret key K_{AP} to encrypt E and then obtains the session key K_{C-AP}. After that, it decrypts G by using K_{C-AP}. Lastly, RS compares two factors userID and timestamp from E and G to decide to grant the access or not.

$$M_5 = \{E, G\}$$

$$G = Enc((userID, timestamp), K_{C-AP})$$

If the result is matched, RS sends back to the client message M6. It helps the client verify the resource service, and acknowledge the key shared with RS.

$$H = Enc\big(timestamp, K_{C-AP}\big)$$

3.2 Fuzzy Extractor

Figure 3 is shown how secure sketch works. General speaking, secure sketch is a technique which allows a noisy input to be reconstructed. It has two parts, SS – Sketch and Rec – Recover. SS procedure takes an input w, then returns a sketch s. To Rec procedure, if its input w' is close to w and there is the helper data s, it can return exactly w. The key point of the secure sketch is the public sketch s does not disclose the biometric information w.

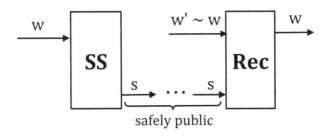

Fig. 3. The secure sketch scheme.

As shown in Fig. 4, fuzzy extractor is a technique which can extract nearly uniform randomness R from its input. This extraction is error-tolerant since R is unchanged even if the input is not the same, as long as the gap between it and the previous one is acceptable. Consequently, R is used as a key to encrypt/decrypt or to authenticate. In other words, fuzzy extractor is counted as biometric tool to authenticate a user using his or her own biometric template as a key.

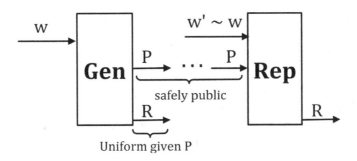

Fig. 4. The fuzzy extractor scheme

In order to preserve high entropy, the authors in [18] employed secure sketch to construct fuzzy extractor. The scheme is demonstrated as Fig. 5. The SS is applied to w to get the sketch s. Along with this procedure, a randomness x and biometrics x are used as input of the strong extractor Ext to obtain the key R. The pair (s, x) is stored as the

helper data P. To reproduce R from w' (close to w), first retrieve P = (s, x), then use
Rec(s, w') to recover w, and lastly Ext(w, x) to have R.

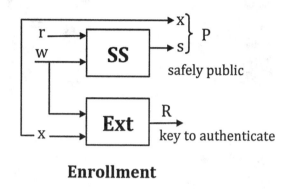

Enrollment

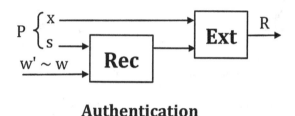

Authentication

Fig. 5. Construction of fuzzy extractors from secure sketches [18]

3.3   Non-invertible Transformation

Non-invertible transformation is one of two types of feature transformation approach.
This technique uses a one-way transformation function to conceal users' biometric
templates in a server side. A one-way function F is "easy to compute" (in polynomial
time) but "hard to invert", and F can be public. The most important thing in non-invertible transformation is how to preserver the similarity of distances among transformed
templates and among original templates (as illustrated in Fig. 6).

Random Orthonormal Projection (ROP) is a technique that utilizes an orthonormal
matrix to project a set of points into other space while preserving the distances between
points. In the categorization of template protection schemes proposed by Jain [17], ROP
belongs to the non-invertible transformation approach. It meets the cancelability requirement by mapping a biometric feature into a secure domain through an orthonormal
matrix. The method to effectively deliver orthonormal matrix was introduced in [16]. It
can be used to replace traditional method of Gram-Schmidt. Given the biometric feature
vector x of size 2n, orthonormal random matrix A of size $2n \times 2n$, random vector b of
size 2n, we have the transformation y = Ax + b.

Two images of same person

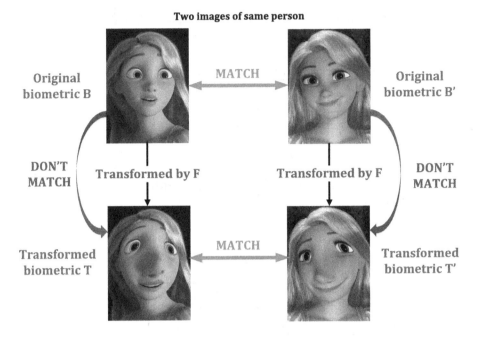

Original
biometric B

MATCH

Original
biometric B'

DON'T
MATCH

Transformed by F

Transformed by F

DON'T
MATCH

Transformed
biometric T

MATCH

Transformed
biometric T'

Fig. 6. Illustration of cancelable biometrics for face [27]

The orthonormal matrix A of size $2n \times 2n$ owns a diagonal which is a set of n orthonormal matrix of size $n \times n$. The other entries of A are zeros. We present the example of matrix A in (1) where the values $\{\theta_1, \theta_2, \dots, \theta_n\}$ are the random numbers in the range $[0:2\pi]$

$$A = \begin{bmatrix} \cos\theta_1 & \sin\theta_1 & 0 & 0 & & 0 & 0 \\ -\sin\theta_1 & \cos\theta_1 & 0 & 0 & & 0 & 0 \\ 0 & 0 & \cos\theta_2 & \sin\theta_2 & & 0 & 0 \\ 0 & 0 & -\sin\theta_2 & \cos\theta_2 & & 0 & 0 \\ 0 & 0 & 0 & 0 & & \cos\theta_n & \sin\theta_n \\ 0 & 0 & 0 & 0 & & -\sin\theta_n & \cos\theta_n \end{bmatrix} \tag{1}$$

By using this technique to produce the orthonormal matrix, there is no need for a complex process such as Gram-Schmidt. Beside its effectiveness in computational complexity, it can also improve the security while guaranteeing intra-class variation. When client is in doubt of his template getting exposed, he only needs to create another orthonormal matrix A to gain a new transformed template.

4   Proposal Protocol

At first, a user generates his biometric key K_{Bio} from his own biometrics B through cancelable biometric key generation process (as illustrated in Fig. 9). The user will register this key to Authentication Server (AS). When the user want to sign in, he regenerates a biometric key and sends it along with some information to AS. Then, AS sends back to the user a packet. That is phase 1. The purpose of this phase is that the user can verify AS from the received packet. In phase 2, the user sends the ticket granted from AS to Ticket Granting Server (TGS), and receives another packet. The purpose of this phase is that TGS can verify the user from the granted ticket. In phase 3, the user extracts the ticket in the packet from TGS to access to Resource Service. The overview of the proposed protocol is demonstrated in Fig. 7.

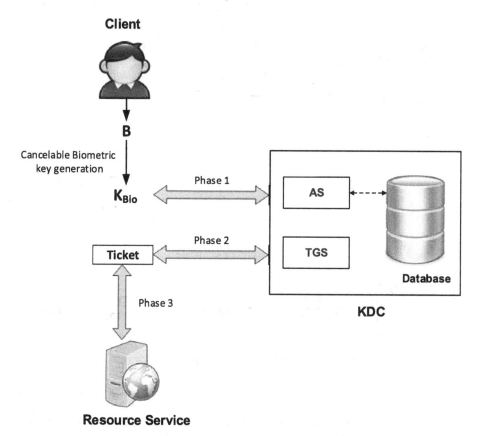

Fig. 7. The overview of the proposed protocol.

4.1 Enrollment Stage

In the enrollment stage, a user provides registration information in a message M01 for Authentication Server – AS (as illustrated in Fig. 8). The Ku_{AS} and Kr_{AS} are the Authentication Server's public key and private key. This pair of keys is used in the next steps to protect the messages between the users and AS, and also to authenticate AS's messages to TGS (Ticket Granting Server). The information in message M01 includes user's mobile number as user's identity, a serial number of the user' device as device's identity, a biometric key K_{bio} which is generated from user's cancelable biometrics by the fuzzy extractor scheme (as illustrated in Fig. 9). The cancelable biometrics B_{TC} resulting from Random Orthonormal Project process of a biometrics B and an orthonormal matrix M can be replaced by another version without changing original biometrics B. If a user has doubt that his biometric key may be compromised, he just needs to choose another random number K'_M to generate another orthonormal matrix M'. With M', the user can recreate a new biometric keyK'_{bio}.

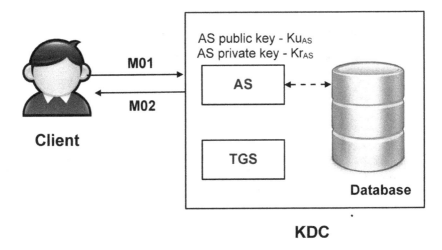

Fig. 8. The enrollment phase.

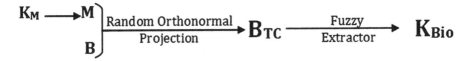

Fig. 9. The cancelable biometric key generation process.

In the registration message, there are also NONCE N_{01} – Number used ONCE and timestamp for defending against replay attack. M01 is encrypted by AS's public key. Once receiving M01, AS employs its private key to decrypt; checks whether there is a duplication of the mobile number; if not, stores the mobile number, the serial number and the biometric key into the database. After that, AS sends M02 back to the client to inform about the successful registration.

$$M_{01} = Enc\big(\big(mobile\_num, serial\_num, K_{bio}, N_{01}, timestamp, is\_enroll\big), Ku_{As}\big)$$

$$M_{02} = Enc\big(\big(N_{01}, timestamp, is\_success\big), Kr_{As}\big)$$

4.2 Authentication Stage

In authentication stage, the goal of the first stage is to verify the authentication server AS. At first, a client sends the message M1 to AS. M1 contains the authentication information such as: the mobile number, the serial number of the device, the number used once N_1 which is generated randomly, and the timestamp. The N_1 is used to verify the message, and the timestamp is used to verify the time. This information is encrypted by the public key of AS Ku_{AS}. Therefore, only can AS decrypt M1 by its private key Kr_{AS}. Getting user's information, AS compares the obtained mobile number and serial number with the stored ones to check if this user belongs to the system or not. Moreover, AS also check the timestamp to be sure this request is a new one. If the information is valid, AS retrieves the biometric key K_{bio} stored in the database to encrypt the message M2, and then sends it back to the client. M2 contains the packet A including the session key (has just been created by AS) between TGS Ks_{TGS} and the client and packet D – the ticket for TGS to verify the client. The communication between the client and AS is demonstrated in Fig. 10.

$$M_1 = Enc\big((mobile\_num, serial\_num, N_1, timestamp), Ku_{As}\big)$$

$$M_2 = Enc\big((A, B), K_{bio}\big)$$

$$A = (timestamp, N_1, Ks_{TGS})$$

$$B = Enc\big((Ks_{TGS}, mobile\_num, serial\_num, N_1, timestamp), K_{TGS}\big)$$

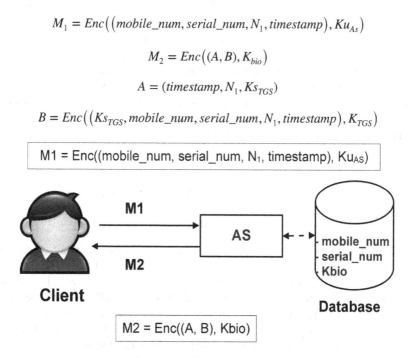

Fig. 10. The first stage of the authentication stage.

Once receiving M2, the user provides his or her biometrics and the number K_M stored in the user's device for the Cancelable Biometric Key Generation program installed directly on the user's device in order to create the biometric key K_{bio}. Right after that, K_{bio} is used to decrypt M2 to obtain the timestamp, N_1 and the session key Ks_{TGS} between the client and TGS. The user compares the timestamp and N_1 to verify the time and his or her request. If they are matched, the user can be sure the AS is valid since only AS has the private key Kr_{AS} to decrypt the message M1 and the biometric key K_{bio} to encrypt the message M2. Thenceforth, the client uses the session key Ks_{TGS} to communicate with TGS.

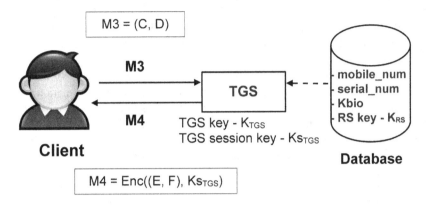

Fig. 11. The second stage of the authentication stage

The second stage is the communication between the client and Ticket Granting Server – TGS. Firstly, the client sends a message M3 to TGS. M3 contains the packet C which includes the user's information for TGS to verify the client, and the resource service's address RS_ID; and packet D which is the ticket AS granted to the client. The communication between the client and TGS is demonstrated in Fig. 11.

$$M_3 = (C, D)$$

$$C = Enc\big((mobile\_num, serial\_num, N_1, timestamp, RS\_ID), Ks_{TGS}\big)$$

$$D = M_2.B$$

Once receiving M3, TGS opens packet D by using its key K_{TGS}, obtains the session key Ks_{TGS} to decrypt C, and the user's information. TGS compares the user's information in the packet D with the information it queries from database, then checks it again with the information in packet C. If they match, the client is successfully authenticated. After verifying client, TGS gets the resource service's address and finds out if this client has the permission to access this service. Subsequently, TGS send the message M4 back to the client. M4 contains the session key Ks_{RS} (has just been created by TGS) between the client and RS and the ticket TGS granted to the client. After that, client employs Ks_{TGS} to decrypt M4 and obtains Ks_{RS}.

$$M_4 = Enc((E, F), Ks_{TGS})$$

$$E = Ks_{RS}$$

$$F = Enc\big((Ks_{RS}, mobile\_num, serial\_num, N_1, timestamp, RS\_ID), K_{RS}\big)$$

The third stage is the communication between the client and the resource service RS. At first, the client sends a message M5 to RS. M5 contains the packet G which includes the user's information for RS to verify the client, and packet H which is the ticket TGS granted to the client (Fig. 12).

$$M_5 = (G, H)$$

$$G = Enc\big((mobile\_num, serial\_num, N_1, timestamp), Ks_{RS}\big)$$

$$H = M_4.F$$

$$M_6 = Enc\big((mobile\_num, serial\_num, N_1 - 1, timestamp), Ks_{RS}\big)$$

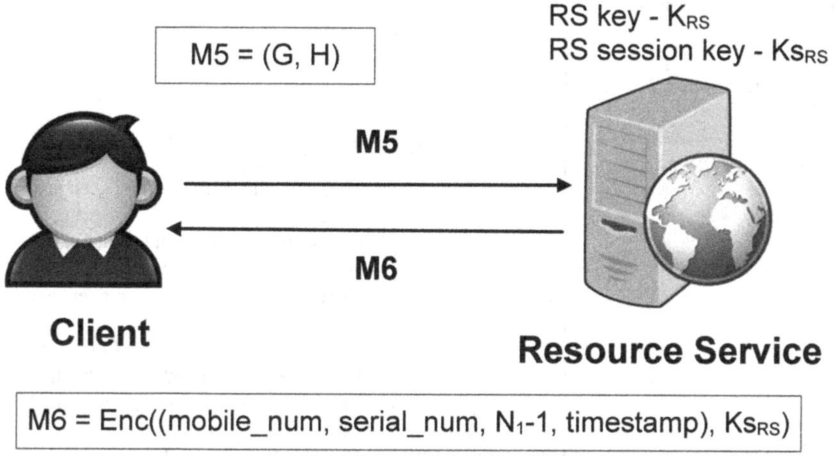

Fig. 12. The third stage of the authentication stage

RS decrypts the packet H by its key K_{RS}, and gets the session key shared with the client Ks_{RS}, then uses the newly obtained key to decrypt the packet G. RS compares the user's information in the packet H and in the packet G to ensure this is the valid user. After that, RS sends the message M6 back to the client in order to help the client verify the resource service. Instead of sending N_1, RS changes the value a little bit to prevent replay attack since M6 and M5.G are almost the same. The whole protocol can be summarized in the Fig. 13.

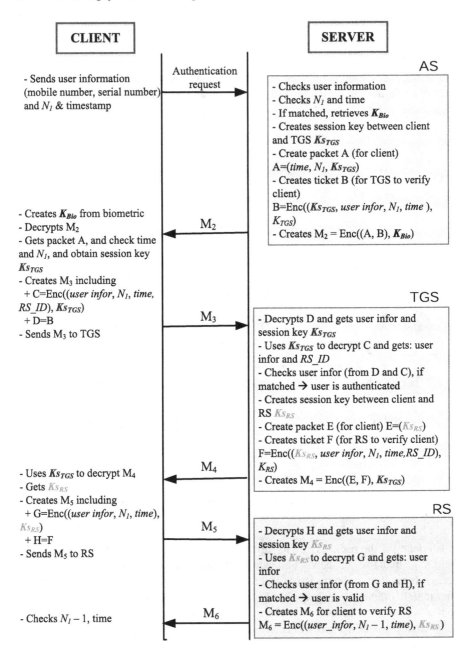

Fig. 13. Summary of the proposed protocol.

5   Security Analysis

In the first message of authentication phase, a client employs the asymmetric cryptosystem and the key length is 2018 bits. Therefore, the security of this message is relatively high. The message M2 is encrypted by the biometric key. From the message M3 to M6, they are all protected by the session keys generated by KDC. The life time of the session keys is very short; hence, if attacks are successful to discover these keys, it doesn't matter since these keys are expired. The weakest point of this system is the message M2 because the biometric key is probably unchanged through the user's lifetime. However, it is hard to attack this key since it is generated from the fuzzy extractor scheme which guarantees the high entropy for the key. Moreover, the key length is 128 bits, it means an attack may need $2^{128} = 340282366920938463463374607431768211456$ probabilities to figure out the right key. In the worst case, even if the biometric key is revealed, it is impossible for an attack to infer the user's biometrics for the high security of the fuzzy extractor scheme. In the case the user doubts that the biometric key is leaked, he or she can alter the random parameter K_M in order to generate another orthonormal matrix M; then the user can get new biometric key from the cancelable biometric key generation process (as illustrated in Fig. 9). Therefore, it can be confirmed that the cancelability requirement is guaranteed in this protocol. In other words, the user's biometrics is strongly protected by fuzzy extractor and non-invertible transformation techniques. The original biometrics is hidden behind the cancelable biometrics. Though attackers or even server obtain the cancelable version, it is hard to turn it back to original form.

For replay attack, it happens when attackers reuse old information to impersonate either client or server with the aim to deceive the other side. In this protocol, this kind of attack is prevented by using timestamp to verify the time and the number used once - NONCE to verify the request. If the timestamp and NONCE doesn't match, the server or even the client can end the session.

The protocol is also immune from man-in-the-middle attack, since it presents mutual authentication requirement, not only does it requires the server to authenticate its right client but also enables the client to perform its own process to confirm requested server. The attack cannot edit the messages transmitted through the network without being detected by server or client.

The protocol supports multi-factor authentication to enhance security of the entire system. It indicates that the authenticity of the client needs following factors: the mobile number, the serial number of the device, and the user's biometrics. If an attacker cannot possess both three factors, he cannot impersonate the user.

Inside attack is also mentioned in this topic. This type of attack happens when the administrator of authentication server exploits user's data stored in the database to legalize his authentication process on behalf of the user. By employing non-invertible transformation and fuzzy extractor, we do not need to store original biometrics into the database server. Therefore, server has no chance to extract the private information of its clients from the biometrics. However, server still can take advantages of the user's data stored into the database in order to connect to the resource services on behalf of the users. In this protocol, the authentication server is divided into AS (Authentication server) and TGS (Ticket grant server). Each server has its own data and function. AS stores users' information and the biometric key; and plays a role of verifying the requesting users. TGS has responsibility for granting ticket to access resource service. TGS does not know the user's biometrics, but it can check the user's authenticity by comparing the information declared by user and the information issued by AS. All these processes attempt to detect the attacks from outside. If AS is an attacker, TGS has nothing to detect this kind of security leak. In brief, our proposal can protect the users' original biometrics against the curiosity of the authentication server; however, it cannot resist the attack from AS happening when AS tries to impersonate its users and deceive both TGS and RS. Strictly speaking, it is the natural setback of Kerberos protocol. TGS and RS have to trust AS and AS is considered as the trusted party in this protocol.

6   Conclusion and Future Works

In this paper, the biometric Kerberos based authentication protocol has been proposed. This protocol relies on Kerberos scheme so it guarantees the mutual authentication between client and server. It also makes the protocol immune from some main attacks over an insecure network. The remarkable point of this work is the embedment fuzzy extractor and non-invertible transformation in the Kerberos scheme to protect a user's biometrics. Biometrics can be concealed into an authentication key. Attacks cannot infer a user's biometrics from this key. And even if, the key is compromised, the user can also generate another biometric key without changing his biometrics. Therefore, the protocol not only guarantees the security but also supports the cancelability property.

In future, we have planned to enhance the remote biometric based protocol. We have aimed at solving the security problems of the untrusted servers. The authentication server cannot take advantages its privileges to obtain users' authentication information and impersonate users. Moreover, we intend to reduce the computational time at client side. Users do not need to perform all processes of creating secure version of their biometrics (in this case, these processes are non-invertible transformation and fuzzy extractor). Authentication server will take the part of responsibility of computing secure biometric template without discovering the original. These requirements can be satisfied by applying the random shares paradigm concept in secure multi-party computation. How to embed this concept in the remote biometric based authentication protocol is our future works.

References

1. Jain, A.K., Ross, A.: Multibiometric systems. Commun. ACM **47**(1), 34–40 (2004)
2. Rathgeb, C., Uhl, A.: A survey on biometric cryptosystems and cancelable biometrics. EURASIP J. Inf. Secur. **2011**(1), 1–25 (2011)
3. Upmanyu, M., et al.: Blind authentication: a secure crypto-biometric verification protocol. Trans. Inf. Forensics Secur. IEEE **5**(2), 255–268 (2010)
4. Nguyen, T.A.T., Dang, T.K.: Combining fuzzy extractor in biometric-kerberos based authentication protocol. In: International Conference on Advanced Computing and Applications, pp. 1–6. IEEE, Ho Chi Minh (2015)

5. Xi, K., et al.: A fingerprint based bio-cryptographic security protocol designed for client/server authentication in mobile computing environment. Secur. Commun. Netw. **4**(5), 487–499 (2011)
6. Nguyen, T.H.L., Nguyen, T.T.H.: An approach to protect private key using fingerprint biometric encryption key in BioPKI based security system. In: The 10th International Conference on Control, Automation, Robotics and Vision, ICARCV (2008)
7. Lifang, W., Songlong, Y.: A face based fuzzy vault scheme for secure online authentication. In: Second International Symposium on Data, Privacy and E-Commerce (ISDPE) (2010)
8. Al-Assam, H., Rashid, R., Jassim, S.: Combining steganography and biometric cryptosystems for secure mutual authentication and key exchange. In: The 8th International Conference for Internet Technology and Secured Transactions, ICITST 2013 (2013)
9. Johnson, R.C., Scheirer, W.J., Boul, T.E.: Secure voice-based authentication for mobile devices: vaulted voice verification (2013)
10. Jonsson, E.: Co-Authentication - a probabilistic approach to authentication, in computer science and engineering. Technical University of Denmark, DTU: Informatics and Mathematical Modelling, Technical University of Denmark, DTU, p. 135 (2007)
11. Wang, F., Han, J.: Multimodal biometric authentication based on score level fusion using support vector machine. Opto-Electron. Rev. **17**(1), 59–64 (2009)
12. Peng, J., et al.: Multimodal biometric authentication based on score level fusion of finger biometrics. Optik-Int. J. Light Electron. Opt. **125**(23), 6891–6897 (2014)
13. Vasuhi, S., et al.: An efficient multi-modal biometric person authentication system using fuzzy logic. In: 2010 Second International Conference on Advanced Computing (ICoAC) (2010)
14. Le, T.T.B., Dang, T.K., Truong, Q.C., Nguyen, T.A.T.: Protecting biometric features by periodic function-based transformation and fuzzy vault. In: Hameurlain, A., Küng, J., Wagner, R., Dang, T.K., Thoai, N. (eds.) TLDKS XVI. LNCS, vol. 8960, pp. 57–70. Springer, Heidelberg (2014). doi:10.1007/978-3-662-45947-8_5
15. Huynh, V.Q.P., et al.: A combination of ANN and secure sketch for generating strong biometric key. J. Sci. Technol. Vietnamese Acad. Sci. Technol. **51**(4B), 30–39 (2013)
16. Al-Assam, H., Sellahewa, H., Jassim, S.: A lightweight approach for biometric template protection. In: Proceedings of SPIE (2009)
17. Jain, A.K., Nandakumar, K., Nagar, A.: Biometric template security. EURASIP J. Adv. Signal Process. **2008**, 1–17 (2008)
18. Dodis, Y., et al.: Fuzzy extractors: how to generate strong keys from biometrics and other noisy data. SIAM J. Comput. **38**(1), 97–139 (2008)
19. Juels, A., Wattenberg, M.: A fuzzy commitment scheme. In: Proceedings of the 6th ACM Conference on Computer and Communications Security, pp. 28–36. ACM: Kent Ridge Digital Labs, Singapore (1999)
20. Ratha, N.K., et al.: Generating cancelable fingerprint templates. IEEE Trans. Pattern Anal. Mach. Intell. **29**(4), 561–572 (2007)
21. Nguyen, T.A.T., Nguyen, D.T., Dang, T.K.: A multi-factor biometric based remote authentication using fuzzy commitment and non-invertible transformation. In: Khalil, I., et al. (eds.) Proceedings of Information and Communication Technology: Third IFIP TC 5/8 International Conference, ICT-EurAsia 2015, and 9th IFIP WG 8.9 Working Conference, CONFENIS 2015, Held as Part of WCC 2015, Daejeon, Korea, 4–7 October 2015, pp. 77–88. Springer, Cham (2015)
22. Failla, P., Sutcu, Y., Barni, M.: eSketch: a privacy-preserving fuzzy commitment scheme for authentication using encrypted biometrics. In: Proceedings of the 12th ACM Workshop on Multimedia and Security, pp. 241–246. ACM, Roma (2010)

23. Lee, C.-C., Hsu, C.-W.: A secure biometric-based remote user authentication with key agreement scheme using extended chaotic maps. Nonlinear Dyn. **71**(1), 201–211 (2013)
24. Zhang, M., Zhang, J., Zhang, Y.: Remote three-factor authentication scheme based on fuzzy extractors. Secur. Commun. Netw. **8**(4), 682–693 (2015)
25. Needham, R.M., Schroeder, M.D.: Using encryption for authentication in large networks of computers. Commun. ACM **21**(12), 993–999 (1978)
26. Kohl, J.T., Neuman, B.C.: The evolution of the Kerberos authentication service. IEEE Computer Society Press, Los Alamitos (1994)
27. Ratha, N., et al.: Privacy enhancements for inexact biometric templates. In: Tuyls, P., Skoric, B., Kevenaar, T. (eds) Security with Noisy Data: On Private Biometrics, Secure Key Storage and Anti-Counterfeiting, pp. 153–168. Springer, London (2007)

Parallel Learning of Local SVM Algorithms for Classifying Large Datasets

Thanh-Nghi Do[1,2(✉)] and François Poulet[3]

[1] College of Information Technology, Can Tho University,
Can Tho 92100, Vietnam
`dtnghi@cit.ctu.edu.vn`
[2] UMI UMMISCO 209, IRD/UPMC, Paris, France
[3] University of Rennes I - IRISA,
Campus de Beaulieu, 35042 Rennes Cedex, France
`francois.poulet@irisa.fr`

Abstract. We propose new parallel learning algorithms of local support vector machines (local SVMs) for effectively non-linear classification of large datasets. The algorithms of local SVMs perform the training task of large datasets with two main steps. The first one is to partition the full dataset into k subsets of data, and then the second one is to learn non-linear SVMs from k subsets to locally classify them in parallel way on multi-core computers. The k local SVMs algorithm (kSVM) uses $k$means clustering algorithm to partition the data into k clusters, then constructs in parallel non-linear SVM models to classify data clusters locally. The decision tree with labeling support vector machines (tSVM) uses C4.5 decision tree algorithm to split the full dataset into terminal-nodes, and then it learns in parallel local SVM models for classifying impurity terminal-nodes with mixture of labels. The krSVM algorithm is to train random ensemble of kSVM. The numerical test results on 4 datasets from UCI repository, 3 benchmarks of handwritten letters recognition and a color image collection of one-thousand small objects show that our proposed algorithms of local SVMs (kSVM, tSVM, krSVM) are efficient compared to the standard SVM (LibSVM) in terms of training time and accuracy for dealing with large datasets.

Keywords: Support vector machines · Local support vector machines · Large-scale non-linear classification

1 Introduction

In recent years, the size of data stored electronically is constantly growing due to the increasing number of internet users, more and more mobile access to internet. For example, there are 1.04 billion daily active users on Facebook. 600,000 h of video are uploaded every day on YouTube while 46,000 years are viewed at the same time. And they are not the most visited web sites (Amazon.com and Yahoo! are on top of them). As almost all the mobile phone can take photos,

© Springer-Verlag GmbH Germany 2017
A. Hameurlain et al. (Eds.): TLDKS XXXI, LNCS 10140, pp. 67–93, 2017.
DOI: 10.1007/978-3-662-54173-9_4

it is estimated that at least 2 trillions photos will be shared on various web sites this year! There are 310 millions people who use Twitter and 600 millions who use Weibo (the "Chinese Twitter"). So there are more and more Internet users generating more and more data. Furthermore the size of the data stored is increasing too, photos have always higher resolution, videos are available in HD with sound in Dolby 5.1, text messages are replaced by voice messages... With such a huge amount of data, it is a big priority to have classification algorithms able to help one in order to find what he is looking for. That is why we present parallel learning algorithms of local Support Vector Machines (SVM) for the classification of very large datasets.

-    The SVM algorithm proposed by [1] and kernel-based methods have shown practical relevance for classification, regression and novelty detection. Successful applications of SVMs have been reported for various fields like face identification, text categorization and bioinformatics [2]. They become increasingly popular data analysis tools. In spite of the prominent properties of SVM, they are not favorable to deal with the challenge of large datasets. SVM solutions are obtained from quadratic programming (QP), so that the computational cost of a SVM approach [3] is at least square of the number of training datapoints and the memory requirement makes SVM impractical. There is a need to scale up learning algorithms to handle massive datasets on personal computers (PCs).

In this extended version of [4,5], we propose new parallel learning algorithms of local SVMs for effective non-linear classification of large datasets. Instead of building a global SVM model, as done by the classical algorithm which is very difficult to deal with large datasets, our algorithms of local SVMs construct an ensemble of local ones that are easily trained by the standard SVM algorithms. The algorithms of local SVMs perform the training task with two main steps. Firstly, the algorithms of local SVMs partition the full dataset into k subsets of data, then the algorithms learn non-linear SVMs from k subsets to locally classify them in the parallel way on multi-core computers. The k local SVMs algorithm (kSVM) uses $k$means clustering algorithm [6] to partition the data into k clusters, and then it constructs in parallel non-linear SVM models to locally classify data clusters. The decision tree with labelling support vector machines (tSVM) uses C4.5 decision tree algorithm [7] to split the full dataset into terminal-nodes, and then it learns in parallel local SVM models for classifying impurity terminal-nodes with mixture of labels. The krSVM algorithm is to train random ensemble of kSVM. The experimental results on 4 datasets from UCI repository [8], 3 benchmarks of handwritten letters recognition [9], MNIST [10,11] and a color image collection of one-thousand small objects, ALOI [12] show that our proposed algorithms (kSVM, tSVM, krSVM) are faster than the standard SVM (LibSVM [13]) in the non-linear classification of large datasets while maintaining high classification accuracy.

The paper is organized as follows. Section 2 briefly introduces the SVM algorithm. Section 3 presents our proposed parallel algorithm of random local SVM for the non-linear classification of large datasets. Section 4 shows the experimental results. Section 5 discusses about related works. We then conclude in Sect. 6.

2   Support Vector Machines

2.1   Support Vector Machines for Binary Classification Problems

Let us consider a linear binary classification task, as depicted in Fig. 1, with m datapoints x_i ($i = 1, \ldots, m$) in the n-dimensional input space R^n, having corresponding labels $y_i = \pm 1$. For this problem, the SVM algorithms [1] try to find the best separating plane (denoted by the normal vector $w \in R^n$ and the scalar $b \in R$), i.e. furthest from both class $+1$ and class -1. It can simply maximize the distance or the margin between the supporting planes for each class ($x.w - b = +1$ for class $+1$, $x.w - b = -1$ for class -1)[1]. The margin between these supporting planes is $2/\|w\|$ (where $\|w\|$ is the 2-norm of the vector w). Any point x_i falling on the wrong side of its supporting plane is considered to be an error, denoted by z_i ($z_i \geq 0$)[2]. Therefore, SVM has to simultaneously maximize the margin and minimize the error. The standard SVMs pursue these goals with the quadratic programming (1).

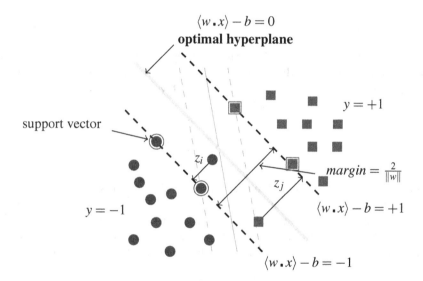

Fig. 1. Linear separation of the datapoints into two classes

$$min\ \Psi(w, b, z) = (1/2)\|w\|^2 + C\sum_{i=1}^{m} z_i$$

(1)

$$s.t. \begin{cases} y_i(w.x_i - b) + z_i \geq 1 \\ z_i \geq 0 \quad \forall i = 1, 2, ..., m \end{cases}$$

---

[1] The inner dot product of two vectors x and y is denoted by $x.y$.

[2] If the point x_i falling on the right side of its supporting plane then its corresponding $z_i = 0$.

where the positive constant C is used to tune errors and margin size.

The plane (w, b) is obtained by solving the quadratic programming (1). Then, the classification function of a new datapoint x based on the plane is:

$$predict(x) = sign(w.x - b) \tag{2}$$

With the mathematical programming concept of duality, the primal quadratic programming problem in (1) is reformulated in the Lagrangian dual one (3) as follows:

$$min\ \Phi(\alpha) = (1/2) \sum_{i=1}^{m} \sum_{j=1}^{m} y_i y_j \alpha_i \alpha_j K\langle x_i, x_j \rangle - \sum_{i=1}^{m} \alpha_i$$

$$s.t. \begin{cases} \sum_{i=1}^{m} y_i \alpha_i = 0 \\ 0 \leq \alpha_i \leq C \quad \forall i = 1, 2, ..., m \end{cases} \tag{3}$$

where C is a positive constant used to tune the margin and the error and a linear kernel function $K\langle x_i, x_j \rangle = \langle x_i.x_j \rangle$.

The support vectors (for which the Lagrange multipliers $\alpha_i > 0$) are given by the solution of the quadratic program (3), and then, the separating surface and the scalar b are determined by the support vectors. The classification of a new data point x based on the SVM model is as follows:

$$predict(x) = sign(\sum_{i=1}^{\#SV} y_i \alpha_i K\langle x, x_i \rangle - b) \tag{4}$$

Variations on SVM algorithms use different classification functions [14]. No algorithmic changes are required from the usual kernel function $K\langle x_i, x_j \rangle$ as a linear inner product, $K\langle x_i, x_j \rangle = \langle x_i.x_j \rangle$ other than the modification of the kernel function evaluation. We can get different support vector classification models. There are two other popular non-linear kernel functions as follows:

– a polynomial function of degree d: $K\langle x_i, x_j \rangle = (\langle x_i.x_j \rangle + 1)^d$
– a RBF (Radial Basis Function): $K\langle x_i, x_j \rangle = e^{-\gamma \|x_i - x_j\|^2}$

SVMs are accurate models with practical relevance for classification, regression and novelty detection. Successful applications of SVMs have been reported for such various fields including facial recognition, text categorization and bioinformatics [2].

2.2 Support Vector Machines for Multi-class Problems

The original SVM algorithms are only able to deal with two-class problems. There are several extensions of a two-class SVM solver for multi-class (c classes, $c \geq 3$) classification tasks. The state-of-the-art multi-class SVMs are categorized into two types of approaches. The first one is to consider the multi-class problem in an optimization problem [15,16]. The second one is to decompose multi-class into a series of binary SVMs, including one-versus-all [1], one-versus-one [17], Decision Directed Acyclic Graph [18] and hierarchical methods for multi-class SVM [19–21] (hierarchically partitioning the data into two subsets).

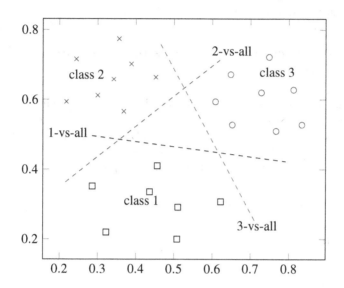

Fig. 2. Multi-class SVM (One-Versus-All)

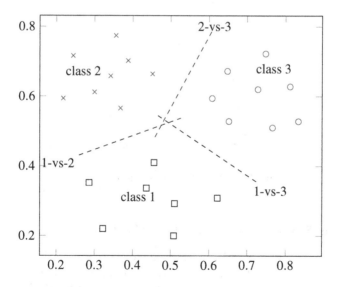

Fig. 3. Multi-class SVM (One-Versus-One)

In practice, the most popular methods are One-Versus-All (ref. LIBLINEAR [22]), One-Versus-One (ref. LibSVM [13]) and are due to their simplicity. The One-Versus-All strategy builds c different binary SVM models where the ith one separates the ith class from the rest, illustrated in Fig. 2. The One-Versus-One strategy constructs $c(c1)/2$ binary SVM models for all the binary pairwise combinations of the c classes, illustrated in Fig. 3. The class is then predicted with the largest distance vote.

3   Parallel Learning Algorithms of Local Support Vector Machines

The study in [3] illustrated that the computational cost requirements of the SVM solutions in (1) or (3) are at least $O(m^2)$ (where m is the number of training datapoints), making standard SVM intractable for large datasets. Learning a global SVM model on the full massive dataset is challenge due to the very high computational cost and the very large memory requirement.

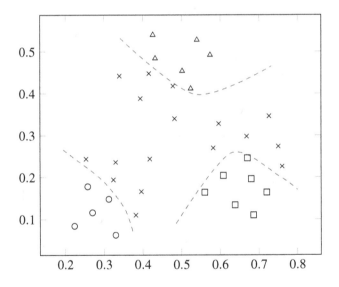

Fig. 4. Global SVM classification model

3.1 k Local Support Vector Machines (kSVM)

Instead of learning a global SVM model, as done by the classical algorithm which is very difficult to deal with large datasets, our proposed local SVMs algorithms learn local SVMs that are easily trained by the standard SVM algorithms. The local SVMs perform the classification task on large datasets into two main steps. The first one is to partition the full dataset into k subsets of data and then the second one is to learn a non-linear SVM in each subset of data to classify the data locally in parallel way on multi-core computers.

Learning k Local SVMs

Our kSVM algorithm proposed in [4] uses a clustering algorithm to partition the full datasets into k clusters and then it is easy to learn a local non-linear SVM (denoted by $lSVM_i$ with the hyper-parameters θ used to train a non-linear SVM model $lSVM_i$) in each cluster (denoted by D_i) to classify the data locally. Figures 4 and 5 show the comparison between a global SVM model and 3 local SVM models, using a non-linear RBF kernel function with $\gamma = 10$ and a positive

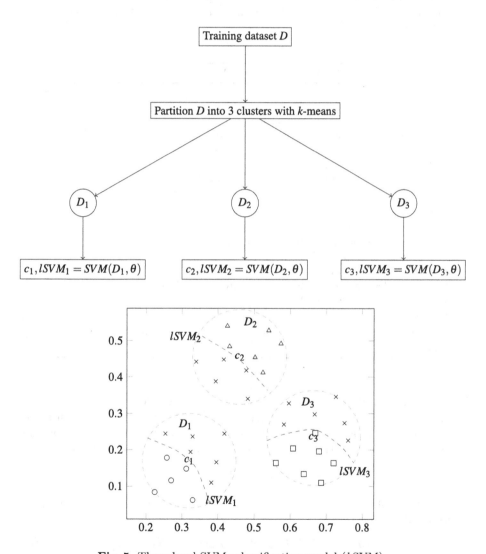

Fig. 5. Three local SVMs classification model (kSVM)

constant $C = 10^6$ (i.e. the hyper-parameters $\theta = \{\gamma, C\}$). In practice, k-means algorithm [6] is the most widely used partitional clustering algorithm because it is simple, easily understandable, and reasonably scalable [23]. Therefore, we propose to use k-means algorithm to partition the full dataset into k clusters and the standard SVM (e.g. LibSVM [13]) to learn k local SVMs.

Furthermore, the kSVM algorithm learns independently k local models from k clusters. This is a nice property for parallel learning. The parallel kSVM does take into account the benefits of high performance computing, e.g. multi-core computers or grids. The simplest development of the parallel kSVM algorithm is based on the shared memory multiprocessing programming model OpenMP [24] on multi-core computers. The parallel training of kSVM is described in Algorithm 1.

Algorithm 1. Local SVM algorithm (kSVM)

---

    input :

        training dataset D

        number of local models k

        hyper-parameter of RBF kernel function γ

        C for tuning margin and errors of SVMs

    output:

        k local support vector machines models

1 **begin**

2   /*k-means performs the data clustering on D;*/

3   creating k clusters denoted by $D_1, D_2, \ldots, D_k$ and

4   their corresponding centers $c_1, c_2, \ldots, c_k$

5   #pragma omp parallel for

6   **for** $i \leftarrow 1$ **to** k **do**

7   /*learning a local SVM model from D_i;*/

8        $lSVM_i = SVM(D_i, \gamma, C)$

9   **end**

10   return $kSVM - model = \{(c_1, lSVM_1), (c_2, lSVM_2), \ldots, (c_k, lSVM_k)\}$

11 **end**

Prediction of a New Individual Using kSVM Model

The $kSVM - model = \{(c_1, lSVM_1), (c_2, lSVM_2), \ldots, (c_k, lSVM_k)\}$ is used to predict the class of a new individual x as follows. The first step is to find the closest cluster based on the distance between x and the cluster centers:

$$c_{NN} = \arg\min_c \; distance(x, c) \tag{5}$$

And then, the class of x is predicted by the local SVM model $lSVM_{NN}$ (corresponding to c_{NN}):

$$predict(x, kSVMmodel) = predict(x, lSVM_{NN}) \tag{6}$$

Performance Analysis of kSVM

Algorithmic Complexity. Let us now examine the complexity of building k local SVM models with the parallel kSVM algorithm. The full dataset with m individuals is partitioned into k balanced clusters (the cluster size is about $\frac{m}{k}$). The training complexity of a local SVM is $O((\frac{m}{k})^2)$. Therefore, the complexity of parallel training k local SVM models on a P-core processor is $O(\frac{k}{P}(\frac{m}{k})^2) = O(\frac{m^2}{kP})$. This complexity analysis illustrates that parallel learning k local SVM models in the kSVM algorithm [3] is kP times faster than building a global SVM model (the complexity is at least $O(m^2)$).

Generalization Capacity. Let us turn back to Theorem 5.2 proposed by Vapnik [25].

Theorem 5.2 ([25] p. 139). If training sets containing m examples are separated by the maximal margin hyperplanes, the expectation (over training sets) of the probability of test error is bounded by the expectation of the minimum of three values: the ratio $\frac{sv}{m}$, where sv is the number of support vectors, the ratio $\frac{1}{m}\frac{R^2}{\Delta}$, where R is the radius of the sphere containing the data and Δ is the value of the margin, and the ratio $\frac{n}{m}$, where n is the dimensionality of the input space:

$$EP_{error} \leq E\left\{min\left(\frac{sv}{m}, \frac{1}{m}\left[\frac{R^2}{\Delta}\right], \frac{n}{m}\right)\right\} \tag{7}$$

Theorem 5.2 illustrates that the maximal margin hyperplane found by the minimization of $\left[\frac{R^2}{\Delta}\right]$ can generalize well. It means that the generalization ability of the large margin hyperplane is high.

In the kSVM, the full dataset with m datapoints is partitioned into k clusters (the cluster size m_k is about $\frac{m}{k}$). Here, the index notation k is used to present m in the context of the cluster (subset). And then the expectation of the probability of test error for a local SVM model (learnt from a cluster) is bounded by:

$$EP_{error} \leq E\left\{min\left(\frac{sv_k}{m_k}, \frac{1}{m_k}\left[\frac{R_k^2}{\Delta_k}\right], \frac{n}{m_k}\right)\right\} \tag{8}$$

The generalization analysis starts with the comparison between the margin size of the global SVM model for the full dataset and the local SVM model learnt from a cluster illustrated in Theorem 1.

Theorem 1. Given a dataset with m datapoints $X = \{x_1, x_2, \ldots, x_m\}$ in the n-dimensional input space R^n, having corresponding labels $Y = \{y_1, y_2, \ldots, y_m\}$ being ± 1, a maximal margin Δ_X hyperplane is to separate furthest from both class $+1$ and class -1, there exists a maximal margin Δ_{X_k} hyperplane for separating a subset of m_k datapoints $X_k \subset X$ into two classes so that the inequality $\Delta_{X_k} \geq \Delta_X$ holds.

---

[3] It must be noted that the complexity of the kSVM approach does not include the k-means clustering used to partition the full dataset. But this step requires insignificant time compared with the quadratic programming solution.

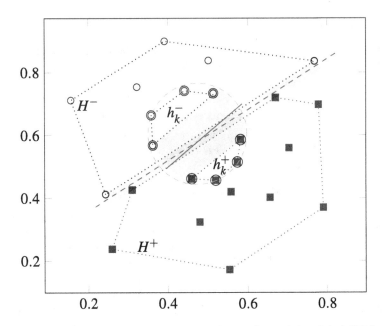

Fig. 6. The comparison of the maximal margin hyperplane of the global SVM and the local SVM

Proof. We remark that the maximal margin Δ_X hyperplane can be seen as the minimum distance between two convex hulls, $H+$ of the positive class P and $H-$ of the negative class N (the farthest distance between the two classes, illustrated in Fig. 6). For subset $X_k \subset X$ containing the subset of the positive class $P_k \subset P$ and the subset of the negative class $N_k \subset N$, it leads to the reduced convex hull h_k+ of $H+$ for the positive class and the reduced convex hull h_k- of $H-$ for the negative class. And then the minimum distance between h_k+ and h_k- can not be smaller than between $H+$ and $H-$. It means that the maximal margin Δ_{X_k} hyperplane for X_k is larger than the maximal margin Δ_X one for fullset X.

The classification performance of a local SVM model in the kSVM is studied in term $\frac{1}{m_k} \left[\frac{R_k^2}{\Delta_k} \right]$ of Eq. 8. In the comparison with the global SVM constructed for the full dataset X, a local SVM model using a subset $X_k \subset X$ of m_k datapoints can guarantee the classification performance because there exists a compromise between the locality (the subset size, i.e. $R_k \leq R$ and $m_k \leq m$) and the generalized capacity (the margin size, i.e. consequence of Theorem 1 $\Delta_{X_k} \geq \Delta_X$).

The generalization analysis illustrates that the parameter k is used in the kSVM to give a trade-off between the generalisation capacity and the computational cost. In [26–28], Vapnik and his colleague also point out the trade-off between the capacity of the local learning system and the number of available individuals. In the context of k local SVM models, this can be understood as follows:

- If k is large then the kSVM algorithm reduces significant training time (the complexity of kSVM is $O(\frac{m^2}{k})$). And then, the size of a cluster is small; the locality is extreme with a very low generalisation capacity.
- If k is small then the kSVM algorithm reduces insignificant training time. However, the size of a cluster is large; it improves the generalisation capacity.

It leads to set k so that the cluster size is large enough (e.g. 200 proposed by [27]).

3.2 Decision Tree Using Labeling Support Vector Machines (tSVM)

Learning tSVM Model

Our proposed tSVM algorithm uses a decision tree algorithm (e.g. C4.5 [7]) to split the full datasets into k partitions. The recursive splitting process of C4.5 algorithm can be early stopped if the number of individuals of the partition is less than the parameter value $minobj$ (the minimum number of individuals that must exist in a node in order for a split to be tried). For the purity terminal-node in which almost all individuals have the same class, the majority label is assigned without any training local SVM models. For the impurity terminal-nodes with mixture of labels, it is to train non-linear SVM models to locally classify these impurity terminal-nodes in the parallel way on multi-core processors (described in Algorithm 2). Figure 7 shows the tSVM model for classifying the same dataset in Fig. 4, using $minobj = 7$ for early stopping the splitting process of C4.5 and a non-linear RBF kernel function with $\gamma = 10$ and a positive constant $C = 10^6$.

Prediction of a New Individual Using tSVM Model

The classification of a new individual x pushes x down the tree t of the tSVM model from the root to the terminal-node. If x arrives at the purity terminal-node then the class of x is the plurality label of this terminal-node; else (x arrives at the impurity terminal-node) the class of x is predicted by the local SVM model learnt from this terminal-node.

Performance Analysis of tSVM

Suppose that the full dataset with m individuals is split into k balanced terminal-nodes (the terminal-node size is about $\frac{m}{k}$; in other words $minobj = \frac{m}{k}$). The training complexity of a SVM for a terminal-node is $O(minobj^2)$. Therefore, the algorithmic complexity of parallel tSVM on a P-core processor is $O(\frac{k}{P}minobj^2) = O(\frac{k}{P}\frac{m}{k}minobj) = O(\frac{m}{P}minobj)$. This complexity analysis [4] illustrates that the speed-up of parallel learning tSVM on a global SVM is $P\frac{m}{minobj}$ times.

tSVM uses the parameter $minobj$ to give a trade-off between the generalisation capacity and the computational cost.

---

[4] It must be noted that the algorithmic complexity of the tSVM approach does not include the C4.5 algorithm used to split the full dataset because this step requires insignificant time compared with the quadratic programming solution.

Algorithm 2. Local SVM algorithm (*t*SVM)

---

   input :
            training dataset D
            minimum number of individuals *minobj* for a split to be tried
            hyper-parameter of RBF kernel function γ
            C for tuning margin and errors of SVMs
   output:
            *t*SVM model

1 **begin**
2   /*Decision tree algorithm partitions the full dataset D;*/
3   learning a tree t for splitting the full dataset D into k partitions
4   denoted by $D_1, D_2, \ldots, D_k$ (using the early stopped parameter *minobj*)
5   #pragma omp parallel for
6   **for** $i \leftarrow 1$ **to** k **do**
7   **if** D_i *is impurity* **then**
8   /*learning a local SVM model from the impurity terminal-node D_i;*/
9          $lSVM_i = SVM(D_i, \gamma, C)$
10   **else**
11         $lSVM_i$ is assigned the plurality label in D_i without any training
12   **end**
13   **end**
14   return $tSVM - model$ = tree t and $\{lSVM_1, lSVM_2, \ldots, lSVM_k\}$
15 **end**

If *minobj* is small then the *t*SVM algorithm significantly reduces training time. And then, the size of a partition is small, the splitting process of C4.5 gives almost purity partitions. *t*SVM becomes a decision tree using the plurality labeling rules at terminal-nodes without any training local SVMs. The locality is extreme with a very low generalisation capacity.

If *minobj* is large then the *t*SVM algorithm reduces insignificant training time. However, the size of a terminal-node is large; It improves the capacity. *t*SVM is a global SVM when *minobj* reaches m.

It leads to set *minobj* large enough (e.g. 200).

3.3   Ensemble of Random Local SVM Models (*kr*SVM)

The analysis of the trade-off between the generalisation capacity and the computational cost illustrates that the local SVM models tries to speed up the training time of the global SVM model while reducing the generalisation capacity. Due to this problem, we propose to construct the ensemble of random local SVM models to improve the generalisation capacity of the local one. The main idea is based on the random forests proposed by Breiman [29]. The randomization is used for controlling high diversity between local SVM models [5]. It leads to the

---

[5] Two classifiers are diverse if they make different errors on new data points [30].

improvement of the generalisation capacity of the single one local SVM model. The ensemble of random local SVM (krSVM described in Algorithm 3) creates a collection of T random local SVMs (kSVM described in Algorithm 1) from bootstrap samples (sampling with replacement from the original dataset) using a randomly chosen subset of attributes. Furthermore, the krSVM constructs independently T random local SVM models (kSVM). It allows parallelizing the learning task with OpenMP [24] on multi-core computers. Thus, the complexity of parallel learning krSVM on a P-core processor is $O(T\frac{m^2}{kP})$.

The prediction class of a new individual x is the plurality class of the classification results obtained by T kSVM models.

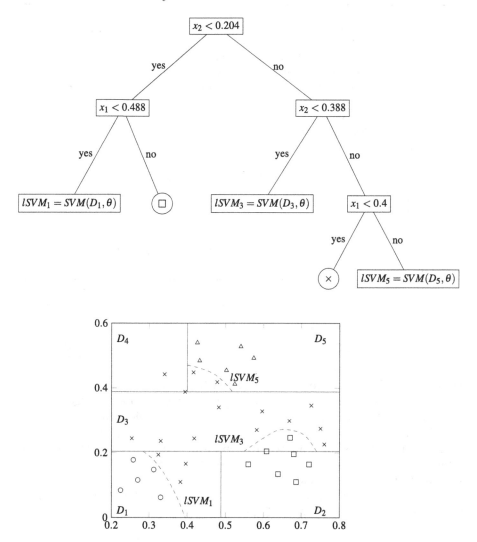

Fig. 7. Decision tree using SVM classifiers at the terminal-nodes (tSVM)

4   Evaluation

We are interested in the performance evaluation of three new parallel learning algorithms of local SVMs for classifying large datasets. Therefore, we conduct the numerical test results to assert their classification performance in terms of correctness and training time, compared to the highly efficient standard SVM library, LibSVM [13].

Algorithm 3. Ensemble of random local SVM (krSVM)

---

    input :

            training dataset D
            number of kSVM models T
            $rdims$ random attributes used in the kSVM model
            k local models in the kSVM model
            hyper-parameter of RBF kernel function γ
            C for tuning margin and errors of SVMs

    output:

            T kSVM models

1 **begin**
2   #pragma omp parallel for
3   **for** $t \leftarrow 1$ **to** T **do**
4   Sampling a bootstrap D_t (train set) from D using $rdims$ random attributes)
5         $kSVM_t = kSVM(D_t, k, \gamma, C)$
6   **end**
7   return $krSVM - model = \{kSVM_1, kSVM_2, \ldots, kSVM_T\}$
8 **end**

4.1   Software Programs

In order to evaluate the classification effectiveness of our local parallel learning algorithms of local SVMs, denoted by kSVM (k local SVMs), tSVM (tree local SVMs), krSVM (ensemble of k random local SVMs), we have implemented them in C/C++, OpenMP [24], using the Automatically Tuned Linear Algebra Software (ATLAS [31]), the free source code of decision tree algorithm C4.5 [7] and LibSVM [13].

The comparative studies are reported in terms of accuracy and training time obtained by our proposed algorithms of local SVMs (kSVM, tSVM and krSVM) and LibSVM (standard global SVM).

All experiments are run on machine Linux Fedora 20, Intel(R) Core i7-4790 CPU, 3.6 GHz, 4 cores and 32 GB RAM.

4.2   Datasets

Experiments are conducted with the 4 datasets collected from UCI repository [8] and the 3 benchmarks of handwritten letters recognition, including USPS [9], MNIST [10], a new benchmark for handwritten character recognition [11] and a color image collection of one-thousand small objects, ALOI [12]. Table 1 presents the description of datasets. The evaluation protocols are illustrated in the last column of Table 1. Datasets are already divided into training set (Trn) and testing set (Tst). We used the training data to build the SVM models. Then, we classified the testing set using the resulting models.

Table 1. Description of datasets

ID	Dataset	Individuals	Attributes	Classes	Evaluation protocol
1	Opt. Rec. of Handwritten Digits	5620	64	10	3832 Trn - 1797 Tst
2	Letter	20000	16	26	13334 Trn - 6666 Tst
3	Isolet	7797	617	26	6238 Trn - 1559 Tst
4	USPS Handwritten Digit	9298	256	10	7291 Trn - 2007 Tst
5	A New Benchmark for HCR	40133	3136	36	36000 Trn - 4133 Tst
6	MNIST	70000	784	10	60000 Trn - 10000 Tst
7	ALOI	108000	128	1000	72000 Trn - 36000 Tst
8	Forest Cover Types	581012	54	7	400000 Trn - 181012 Tst

The three last datasets (MNIST, ALOI, Forest Cover Types) in Table 1 yield large-scale non-linear SVM classification problems. Typically, the training time of a non-linear SVM for Forest Cover Types is at least 23 days [32–36].

4.3   Tuning Parameters

We propose to use RBF kernel type in local SVMs and SVM models because it is general and efficient [37]. We also tried to tune the hyper-parameter γ of RBF kernel (RBF kernel of two individuals x_i, x_j, $K[i,j] = exp(-\gamma\|x_i - x_j\|^2)$) and the cost C (a trade-off between the margin size and the errors) to obtain a good accuracy.

For the parameter k local models (number of clusters) of kSVM and krSVM, we propose to set k so that each cluster has about 1000 individuals. The idea gives a trade-off between the generalization capacity [28] and the computational cost.

For the parameter $minobj$ in tSVM (the minimum number of individuals that must exist in a node in order for a split to be attempted) is set equal to 1000.

Furthermore, our krSVM uses 20 random k local SVM models with the number of random ($rdims$) attributes being one half full set.

Table 2 presents the hyper-parameters of kSVM, tSVM, krSVM and LibSVM in the classification.

Table 2. Hyper-parameters of kSVM, tSVM, krSVM and LibSVM

ID	Datasets	γ	C	k	$minobj$
1	Opt. Rec. of Handwritten Digits	0.0001	100000	10	1000
2	Letter	0.0001	100000	30	1000
3	Isolet	0.0001	100000	10	1000
4	USPS Handwritten Digit	0.0001	100000	10	1000
5	A New Benchmark for HCR	0.001	100000	50	1000
6	MNIST	0.05	100000	100	1000
7	ALOI	0.01	100000	100	1000
8	Forest Cover Types	0.0001	100000	500	1000

4.4   Classification Results

The classification results of kSVM, tSVM, krSVM and LibSVM on the 8 datasets are given in Tables 3 and 4 and Figs. 8, 9, 10, 11 and 12.

As it was expected, our three algorithms of local SVMs (kSVM, tSVM, krSVM) outperform LibSVM in terms of training time. The average training time kSVM, tSVM, krSVM, LibSVM are 44.63 s, 54.85 s, 75.57 s, at least 4646.98 s, respectively. kSVM is the fastest learning algorithm. tSVM holds the rank 2. krSVM is about 1.7 times slower than kSVM. Training time of LibSVM is at least 61 times longer than krSVM.

Table 3. Training time (s)

ID	Datasets	Training time(s)			
		LibSVM	krSVM	kSVM	tSVM
1	Opt. Rec. of Handwritten Digits	0.58	0.54	0.21	0.12
2	Letter	2.87	1.94	0.5	0.42
3	Isolet	8.37	7.14	2.94	3.98
4	USPS Handwritten Digit	5.88	5.32	3.82	4.62
5	A New Benchmark for HCR	107.07	91.72	35.7	95.37
6	MNIST	1531.06	82.26	45.5	124.48
7	ALOI	2400	142.28	44.68	30
8	Forest Cover Types	NA (> 33120)	273.36	223.7	179.84
	Average	NA (> 4646.98)	75.57	44.63125	54.85375

For the 5 first small datasets, the classification results (presented in Tables 3 and 4 and Fig. 8) show that the improvement of local SVM algorithms against LibSVM is slight.

For the 3 last large datasets, the classification performances in Tables 3 and 4 and Figs. 9, 10 and 11 demonstrate that the speed-up in learning of local SVM algorithms on LibSVM is significant.

Table 4. Classification results in terms of accuracy (%)

ID	Datasets	Classification accuracy(%)			
		LibSVM	krSVM	kSVM	tSVM
1	Opt. Rec. of Handwritten Digits	98.33	97.61	97.05	96.99
2	Letter	97.40	97.16	96.14	95.65
3	Isolet	96.47	96.15	95.44	95.38
4	USPS Handwritten Digit	96.86	96.46	95.86	95.02
5	A New Benchmark for HCR	95.14	94.77	92.98	92.72
6	MNIST	98.37	98.71	98.11	98.24
7	ALOI	95.16	95.09	93.33	93.17
8	Forest Cover Types	NA	97.07	97.06	96.73
Average		96.82	96.63	95.75	95.49

For MNIST dataset, krSVM is 18.61 times faster than LibSVM while maintaining a slight correctness improvement (0.34%). kSVM and tSVM perform the classification with very competitive correctness compared to LibSVM but their training time are 33 times and 12 times faster than LibSVM.

ALOI dataset has very large number of classes (1000 classes). LibSVM uses the One-Versus-One strategy, therefore it requires 499500 binary SVM models for training and testing. And then, LibSVM classifies this dataset in 2400 s with 95.13% accuracy. krSVM performs the classification task in 142.28 s with 95.09% accuracy. The speed-up of krSVM on LibSVM is about 16 times without loss of the test correctness. kSVM is about 53 times faster than LibSVM with loss of 1.83% accuracy. tSVM is about 80 times faster than LibSVM with loss of 1.99% accuracy.

Typically, Forest cover type dataset is well-known as a difficult dataset for non-linear SVM [32,33]; LibSVM ran for 23 days without any result. krSVM performs this non-linear classification in 273.36 s with 97.07% accuracy. kSVM takes 223.7 s of the training time with the same accuracy as krSVM. tSVM achieves an accuracy of 96.73 in 179.84 s training time.

The classification results in terms of test correctness (presented in Table 4 and Fig. 12) show that our algorithms of local SVMs give very competitive correctness compared to LibSVM. kSVM, tSVM, krSVM and LibSVM achieve the average accuracy of 95.75%, 95.49%, 96.63%, 96.82%, respectively. The superiority of LibSVM on krSVM corresponds to 0.19%. The classification results show that krSVM has an improvement of 0.88% compared to kSVM and of 1.14% compared to tSVM.

5   Discussion on Related Works

Our proposal is related to large-scale SVM learning algorithms. The improvements of SVM training on very large datasets include effective heuristic methods

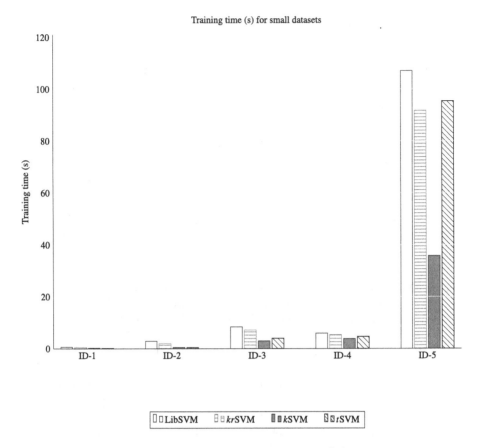

Fig. 8. Comparison of training time on small datasets

in the decomposition of the original quadratic programming into series of small problems [3, 38, 39] and [13].

Mangasarian and his colleagues proposed to modify SVM problems to obtain new formulas, including Lagrangian SVM [40], proximal SVM [41], Newton SVM [42]. The Least Squares SVM proposed by Suykens and Vandewalle [43] changes standard SVM optimization to lead the new efficient SVM solver. And then, these algorithms only require solving a system of linear equations instead of a quadratic programming. This makes training time very short for linear classification tasks. Their extensions proposed by [33, 44–47] aim at improving memory performance for massive datasets by incrementally updating solutions in a growing training set without needing to load the entire dataset into memory at once. More recent [48, 49] proposed the stochastic gradient descent methods for dealing with large scale linear SVM solvers. The parallel and distributed algorithms [45, 47, 50, 51] for the linear classification improve learning performance for large datasets by dividing the problem into sub-problems that execute on large numbers of networked PCs, grid computing, multi-core computers. Parallel SVMs proposed by [52] use GPU

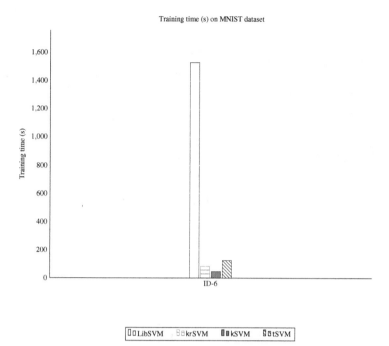

Fig. 9. Comparison of training time on MNIST dataset

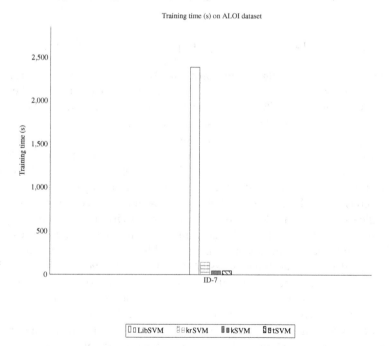

Fig. 10. Comparison of training time on ALOI dataset

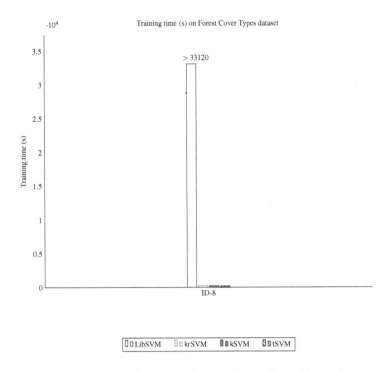

Fig. 11. Comparison of training time on Forest Cover Types dataset

to speed-up training tasks. These algorithms are efficient for linear classification tasks.

Active SVM learning algorithms proposed by [32,53–55] choose interesting datapoint subsets (active sets) to construct models, instead of using the whole dataset. SVM algorithms [33,56–58] use the boosting strategy [59,60] for the linear classification of very large datasets on standard PCs.

Recently, the review paper [61] provides a comprehensive survey on large-scale linear support vector classification. The parallel algorithms of logistic regression and linear SVM using Spark [62] are proposed in [63]. The distributed Newton algorithm of logistic regression [64] is implemented in the Message Passing Interface (MPI). The parallel dual coordinate descent method for linear classification [65] is implemented in multi-core environments using OpenMP. The incremental and decremental algorithms [66] aim at training linear classification of large data that cannot fit in memory. These algorithms are proposed to efficiently deal large-scale linear classification tasks in a very-high-dimensional input space. But the computational cost of a non-linear SVM approach is still impractical. The work in [67] tries to automatically determine which kernel classifiers perform strictly better than linear for a given data set.

Our proposal is in some aspects related to local SVM learning algorithms. The first approach is to classify data in hierarchical strategy. This kind of training algorithm performs the classification task with two main steps. The first

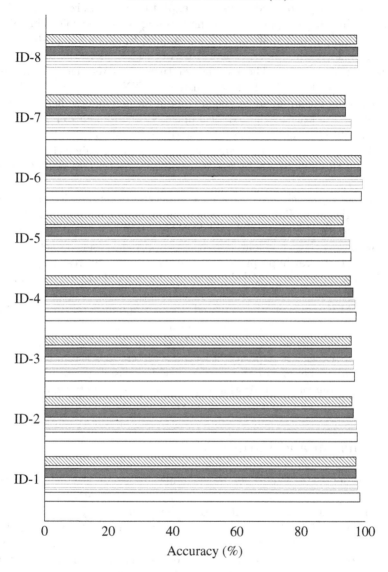

Classification correctness (%)

Fig. 12. Comparison of classification results

one is to cluster the full dataset into homogeneous groups (clusters) and then the second one is to learn the local supervised classification models from clusters. The paper [68] proposed to use the expectation-maximization (EM) clustering algorithm [69] for partitioning the training set into k joint clusters (the EM clustering algorithm makes a soft assignment based on the posterior probabilities [70]); for each cluster, a neural network (NN) is learnt to classify the individuals in the cluster. The parallel mixture of SVMs algorithm proposed by [34] constructs local SVM models instead of NN ones in [68]. CSVM [71] uses k-means algorithm [6] to partition the full dataset into k disjoint clusters; then the algorithm learns weighted local linear SVMs from clusters. More recent DTSVM [36,72] uses the decision tree algorithm [7,73] to split the full dataset into disjoint regions (tree leaves) and then the algorithm builds the local SVMs for classifying the individuals in tree leaves. These algorithms aim at speeding up the learning time.

The second approach is to learn local supervised classification models from k nearest neighbors (kNN) of a new testing individual. First local learning algorithm of Bottou and Vapnik [27] find kNN of a test individual; train a neural network with only these k neighborhoods and apply the resulting network to the test individual. k-local hyperplane and convex distance nearest neighbor algorithms are also proposed in [74]. More recent local SVM algorithms aim to use the different methods for kNN retrieval; including SVM-kNN [75] trying different metrics, ALH [76] using the weighted distance and features, FaLK-SVM [35] speeding up the kNN retrieval with the cover tree [77].

A theoretical analysis for such local algorithms discussed in [26] introduces the trade-off between the capacity of learning system and the number of available individuals. The size of the neighborhoods is used as an additional free parameters to control generalisation capacity against locality of local learning algorithms.

6   Conclusion and Future Works

We have presented new parallel learning algorithms of local support vector machines that achieve high performances (in terms of training time and correctness) for the non-linear classification of large datasets. The training task of the local SVMs in kSVM, tSVM and krSVM algorithms is to partition the full data into k subsets of data and then it constructs a non-linear SVM in each subset of data to locally classify them in parallel way. The numerical test results on 4 datasets from UCI repository, 3 benchmarks of handwritten letters recognition and a color image collection of one-thousand small objects showed that our proposed algorithms are efficient in terms of training time and accuracy compared to the standard SVM. An example of their effectiveness is given with the non-linear classification of Forest Cover Types dataset (having 581012 individuals, 54 attributes) into 7 classes, krSVM classifies this dataset in 273.36 s with 97.07% accuracy. kSVM takes 223.7 s of the training time with the same accuracy as krSVM. tSVM achieves an accuracy of 96.73% in 179.84 s training time.

In the near future, we intend to provide more empirical test on large benchmarks and comparisons with other algorithms. A promising future research aims at automatically tuning the hyper-parameters of local SVMs.

References

1. Vapnik, V.N.: The Nature of Statistical Learning Theory. Springer, New York (1995)
2. Guyon, I.: Web page on SVM applications (1999). http://www.clopinet.com/isabelle/Projects/-SVM/app-list.html
3. Platt, J.: Fast training of support vector machines using sequential minimal optimization. In: Schölkopf, B., Burges, C., Smola, A. (eds.) Advances in Kernel Methods Support Vector Learning, pp. 185–208 (1999)
4. Do, T.-N.: Non-linear classification of massive datasets with a parallel algorithm of local support vector machines. In: Le Thi, H.A., Nguyen, N.T., Do, T.V. (eds.) Advanced Computational Methods for Knowledge Engineering. AISC, vol. 358, pp. 231–241. Springer, Heidelberg (2015). doi:10.1007/978-3-319-17996-4_21
5. Do, T.-N., Poulet, F.: Random local SVMs for classifying large datasets. In: Dang, T.K., Wagner, R., Küng, J., Thoai, N., Takizawa, M., Neuhold, E. (eds.) FDSE 2015. LNCS, vol. 9446, pp. 3–15. Springer, Heidelberg (2015). doi:10.1007/978-3-319-26135-5_1
6. MacQueen, J.: Some methods for classification and analysis of multivariate observations. In: Proceedings of 5th Berkeley Symposium on Mathematical Statistics and Probability, vol. 1, pp. 281–297. University of California Press, Berkeley, January 1967
7. Quinlan, J.R.: C4.5: Programs for Machine Learning. Morgan Kaufmann, San Mateo (1993)
8. Lichman, M.: UCI machine learning repository (2013)
9. LeCun, Y., Boser, B., Denker, J., Henderson, D., Howard, R., Hubbard, W., Jackel, L.: Backpropagation applied to handwritten zip code recognition. Neural Comput. **1**(4), 541–551 (1989)
10. LeCun, Y., Bottou, L., Bengio, Y., Haffner, P.: Gradient-based learning applied to document recognition. Proc. IEEE **86**(11), 2278–2324 (1998)
11. van der Maaten, L.: A new benchmark dataset for handwritten character recognition (2009). http://homepage.tudelft.nl/19j49/Publications_files/characters.zip
12. Geusebroek, J.M., Burghouts, G.J., Smeulders, A.W.M.: The amsterdam library of object images. Intl. J. Comput. Vis. **61**(1), 103–112 (2005)
13. Chang, C.C., Lin, C.J.: LIBSVM : a library for support vector machines. ACM Trans. Intell. Syst. Technol. **2**(27), 1–27 (2011)
14. Cristianini, N., Shawe-Taylor, J.: An Introduction to Support Vector Machines: And Other Kernel-based Learning Methods. Cambridge University Press, New York (2000)
15. Weston, J., Watkins, C.: Support vector machines for multi-class pattern recognition. In: Proceedings of the Seventh European Symposium on Artificial Neural Networks, pp. 219–224 (1999)
16. Guermeur, Y.: VC theory of large margin multi-category classifiers. J. Mach. Learn. Res. **8**, 2551–2594 (2007)
17. Kreßel, U.: Pairwise classification and support vector machines. In: Advances in Kernel Methods: Support Vector Learning, pp. 255–268 (1999)

18. Platt, J., Cristianini, N., Shawe-Taylor, J.: Large margin dags for multiclass classification. Adv. Neural Inf. Process. Syst. **12**, 547–553 (2000)
19. Vural, V., Dy, J.: A hierarchical method for multi-class support vector machines. In: Proceedings of the Twenty-First International Conference on Machine Learning, pp. 831–838 (2004)
20. Benabdeslem, K., Bennani, Y.: Dendogram-based SVM for multi-class classification. J. Comput. Inf. Technol. **14**(4), 283–289 (2006)
21. Do, T.N., Lenca, P., Lallich, S.: Classifying many-class high-dimensional fingerprint datasets using random forest of oblique decision trees. Vietnam J. Comput. Sci. **2**(1), 3–12 (2015)
22. Fan, R., Chang, K., Hsieh, C., Wang, X., Lin, C.: LIBLINEAR: a library for large linear classification. J. Mach. Learn. Res. **9**(4), 1871–1874 (2008)
23. Wu, X., Kumar, V., Quinlan, J.R., Ghosh, J., Yang, Q., Motoda, H., McLachlan, G.J., Ng, A., Liu, B., Philip, S.Y., Zhou, Z.H., Steinbach, M., Hand, D.J., Steinberg, D.: Top 10 algorithms in data mining. Knowl. Inf. Syst. **14**(1), 1–37 (2007)
24. OpenMP Architecture Review Board: OpenMP Application Program Interface Version 3.0 (2008)
25. Vapnik, V.N.: The Nature of Statistical Learning Theory, 2nd edn. Springer, New York (2000)
26. Vapnik, V.: Principles of risk minimization for learning theory. In: Advances in Neural Information Processing Systems 4, NIPS Conference, Denver, Colorado, USA, December 2–5, 1991, pp. 831–838 (1991)
27. Bottou, L., Vapnik, V.: Local learning algorithms. Neural Comput. **4**(6), 888–900 (1992)
28. Vapnik, V., Bottou, L.: Local algorithms for pattern recognition and dependencies estimation. Neural Comput. **5**(6), 893–909 (1993)
29. Breiman, L.: Random forests. Mach. Learn. **45**(1), 5–32 (2001)
30. Dietterich, T.G.: Ensemble methods in machine learning. In: Kittler, J., Roli, F. (eds.) MCS 2000. LNCS, vol. 1857, pp. 1–15. Springer, Heidelberg (2000). doi:10.1007/3-540-45014-9_1
31. Whaley, R., Dongarra, J.: Automatically tuned linear algebra software. In: Ninth SIAM Conference on Parallel Processing for Scientific Computing, CD-ROM Proceedings (1999)
32. Yu, H., Yang, J., Han, J.: Classifying large data sets using SVMs with hierarchical clusters. In: Proceedings of the ACM SIGKDD International Conference on KDD, pp. 306–315. ACM (2003)
33. Do, T.N., Poulet, F.: Towards high dimensional data mining with boosting of PSVM and visualization tools. In: Proceedings of 6th International Conference on Entreprise Information Systems, pp. 36–41(2004)
34. Collobert, R., Bengio, S., Bengio, Y.: A parallel mixture of SVMs for very large scale problems. Neural Comput. **14**(5), 1105–1114 (2002)
35. Segata, N., Blanzieri, E.: Fast and scalable local kernel machines. J. Learn. Res. **11**, 1883–1926 (2010)
36. Chang, F., Guo, C.Y., Lin, X.R., Lu, C.J.: Tree decomposition for large-scale SVM problems. J. Mach. Learn. Res. **11**, 2935–2972 (2010)
37. Lin, C.: A practical guide to support vector classification (2003)
38. Boser, B., Guyon, I., Vapnik, V.: An training algorithm for optimal margin classifiers. In: Proceedings of 5th ACM Annual Workshop on Computational Learning Theory of 5th ACM Annual Workshop on Computational Learning Theory, pp. 144–152. ACM (1992)

39. Osuna, E., Freund, R., Girosi, F.: An improved training algorithm for support vector machines. In: Principe, J., Gile, L., Morgan, N., Wilson, E. (eds.) Neural Networks for Signal Processing VII, pp. 276–285 (1997)

40. Mangasarian, O., Musicant, D.: Lagrangian support vector machines. J. Mach. Learn. Res. **1**, 161–177 (2001)

41. Fung, G., Mangasarian, O.: Proximal support vector classifiers. In: Proceedings of the ACM SIGKDD International Conference on KDD, pp. 77–86. ACM (2001)

42. Mangasarian, O.: A finite Newton method for classification problems. Technical report, pp. 01–11. Data Mining Institute, Computer Sciences Department, University of Wisconsin (2001)

43. Suykens, J., Vandewalle, J.: Least squares support vector machines classifiers. Neural Process. Lett. **9**(3), 293–300 (1999)

44. Do, T.N., Poulet, F.: Incremental SVM and visualization tools for bio-medical data mining. In: Proceedings of Workshop on Data Mining and Text Mining in Bioinformatics, pp. 14–19 (2003)

45. Do, T.N., Poulet, F.: Classifying one billion data with a new distributed SVM algorithm. In: Proceedings of 4th IEEE International Conference on Computer Science, Research, Innovation and Vision for the Future, pp. 59–66. IEEE Press (2006)

46. Fung, G., Mangasarian, O.: Incremental support vector machine classification. In: Proceedings of the 2nd SIAM International Conference on Data Mining (2002)

47. Poulet, F., Do, T.N.: Mining very large datasets with support vector machine algorithms. In: Camp, O., Filipe, J., Hammoudi, S., Piattini, M. (eds.) Enterprise Information Systems V, pp. 177–184. Springer, Amsterdam (2004)

48. Shalev-Shwartz, S., Singer, Y., Srebro, N.: Pegasos: primal estimated sub-gradient solver for SVM. In: Proceedings of the Twenty-Fourth International Conference Machine Learning, pp. 807–814. ACM (2007)

49. Bottou, L., Bousquet, O.: The tradeoffs of large scale learning. In: Platt, J., Koller, D., Singer, Y., Roweis, S. (eds.) Advances in Neural Information Processing Systems, vol. 20, pp. 161–168. NIPS Foundation (2008). http://books.nips.cc

50. Do, T.N.: Parallel multiclass stochastic gradient descent algorithms for classifying million images with very-high-dimensional signatures into thousands classes. Vietnam J. Comput. Sci. **1**(2), 107–115 (2014)

51. Doan, T., Do, T., Poulet, F.: Large scale classifiers for visual classification tasks. Multimedia Tools Appl. **74**(4), 1199–1224 (2015)

52. Do, T.-N., Nguyen, V.-H., Poulet, F.: Speed up SVM algorithm for massive classification tasks. In: Tang, C., Ling, C.X., Zhou, X., Cercone, N.J., Li, X. (eds.) ADMA 2008. LNCS (LNAI), vol. 5139, pp. 147–157. Springer, Heidelberg (2008). doi:10.1007/978-3-540-88192-6_15

53. Do, T.N., Poulet, F.: Mining very large datasets with SVM and visualization. In: Proceedings of 7th International Conference on Entreprise Information Systems, pp. 127–134 (2005)

54. Boley, D., Cao, D.: Training support vector machines using adaptive clustering. In: Berry, M.W., Dayal, U., Kamath, C., Skillicorn, D.B. (eds.) Proceedings of the Fourth SIAM International Conference on Data Mining, Lake Buena Vista, Florida, USA, 22–24 April, 2004, SIAM, pp. 126–137 (2004)

55. Tong, S., Koller, D.: Support vector machine active learning with applications to text classification. In: Proceedings of the 17th International Conference on Machine Learning, pp. 999–1006. ACM (2000)

56. Pavlov, D., Mao, J., Dom, B.: Scaling-up support vector machines using boosting algorithm. In: 15th International Conference on Pattern Recognition, vol. 2, pp. 219–222 (2000)
57. Do, T.N., Le-Thi, H.A.: Classifying large datasets with SVM. In: Proceedings of 4th International Conference on Computational Management Science (2007)
58. Do, T.N., Fekete, J.D.: Large scale classification with support vector machine algorithms. In: Wani, M.A., Kantardzic, M.M., Li, T., Liu, Y., Kurgan, L.A., Ye, J., Ogihara, M., Sagiroglu, S., Chen, X.W., Peterson, L.E., Hafeez, K. (eds.) The Sixth International Conference on Machine Learning and Applications, ICMLA 2007, Cincinnati, Ohio, USA, 13–15 December 2007, pp. 7–12. IEEE Computer Society (2007)
59. Freund, Y., Schapire, R.: A short introduction to boosting. J. Jpn. Soc. Artif. Intell. **14**(5), 771–780 (1999)
60. Breiman, L.: Arcing classifiers. Ann. Stat. **26**(3), 801–849 (1998)
61. Yuan, G., Ho, C., Lin, C.: Recent advances of large-scale linear classification. Proc. IEEE **100**(9), 2584–2603 (2012)
62. Zaharia, M., Chowdhury, M., Franklin, M.J., Shenker, S., Stoica, I.: Spark: cluster computing with working sets. In: Proceedings of the 2nd USENIX Conference on Hot Topics in Cloud Computing, HotCloud 2010, Berkeley, CA, USA, p. 10. USENIX Association (2010)
63. Lin, C., Tsai, C., Lee, C., Lin, C.: Large-scale logistic regression and linear support vector machines using spark. In: 2014 IEEE International Conference on Big Data, Big Data 2014, Washington, DC, USA, 27–30 October, 2014, pp. 519–528 (2014)
64. Zhuang, Y., Chin, W.-S., Juan, Y.-C., Lin, C.-J.: Distributed Newton methods for regularized logistic regression. In: Cao, T., Lim, E.-P., Zhou, Z.-H., Ho, T.-B., Cheung, D., Motoda, H. (eds.) PAKDD 2015. LNCS (LNAI), vol. 9078, pp. 690–703. Springer, Heidelberg (2015). doi:10.1007/978-3-319-18032-8_54
65. Chiang, W., Lee, M., Lin, C.: Parallel dual coordinate descent method for large-scale linear classification in multi-core environments. In: Proceedings of the 22nd ACM SIGKDD International Conference on Knowledge Discovery and Data Mining, San Francisco, CA, USA, August 13–17, 2016, pp. 1485–1494 (2016)
66. Tsai, C., Lin, C., Lin, C.: Incremental and decremental training for linear classification. In: The 20th ACM SIGKDD International Conference on Knowledge Discovery and Data Mining, KDD 2014, New York, NY, USA , 24–27 August, 2014, pp. 343–352 (2014)
67. Huang, H., Lin, C.: Linear and kernel classification: when to use which? In: Proceedings of the SIAM International Conference on Data Mining 2016 (2016)
68. Jacobs, R.A., Jordan, M.I., Nowlan, S.J., Hinton, G.E.: Adaptive mixtures of local experts. Neural Comput. **3**(1), 79–87 (1991)
69. Dempster, A.P., Laird, N.M., Rubin, D.B.: Maximum likelihood from incomplete data via the EM algorithm. J. Roy. Stat. Soc. Ser. B **39**(1), 1–38 (1977)
70. Bishop, C.M.: Pattern Recognition and Machine Learning. Springer, New York (2006)
71. Gu, Q., Han, J.: Clustered support vector machines. In: Proceedings of the Sixteenth International Conference on Artificial Intelligence and Statistics, AISTATS 2013, Scottsdale, AZ, USA, 29 April–1 May, 2013, JMLR Proceedings, vol. 31, pp. 307–315(2013)
72. Chang, F., Liu, C.C.: Decision tree as an accelerator for support vector machines. In: Ding, X. (ed.) Advances in Character Recognition. InTech (2012)
73. Breiman, L., Friedman, J.H., Olshen, R.A., Stone, C.: Classification and Regression Trees. Wadsworth International, Monterey (1984)

74. Vincent, P., Bengio, Y.: K-local hyperplane and convex distance nearest neighbor algorithms. In: Advances in Neural Information Processing Systems, pp. 985–992. The MIT Press (2001)

75. Zhang, H., Berg, A., Maire, M., Malik, J.: SVM-KNN: discriminative nearest neighbor classification for visual category recognition. In: 2006 IEEE Computer Society Conference on Computer Vision and Pattern Recognition, vol. 2, pp. 2126–2136 (2006)

76. Yang, T., Kecman, V.: Adaptive local hyperplane classification. Neurocomputing $71(1315)$, 3001–3004 (2008)

77. Beygelzimer, A., Kakade, S., Langford, J.: Cover trees for nearest neighbor. In: Proceedings of the 23rd International Conference on Machine Learning, pp. 97–104. ACM (2006)

Contractual Specifications of Business Services: Modeling, Formalization and Proximity

Lam-Son Lê[1]([✉]), Trung-Viet Nguyen[1,2],
Thai-Minh Truong[1], and Khuong Nguyen-An[1]

[1] Faculty of Computer Science and Engineering,
HCMC University of Technology, Ho Chi Minh City, Vietnam
lam-son.le@alumni.epfl.ch, {1680963,thaiminh,nakhuong}@hcmut.edu.vn
[2] Faculty of Information Technology,
Can Tho University of Technology, Can Tho City, Vietnam
ntviet@ctuet.edu.vn

Abstract. Business services arguably play a central role in service-based information systems as they would fill in the gap between the technicality of Service-Oriented Architecture and the business aspects captured in Enterprise Architecture. Business services have distinctive features that are not typically observed in Web services, e.g. significant portions of the functionality of business services might be executed in a human-mediated fashion. The representation of business services requires that we view human activity and human-mediated functionality through the lens of computing and systems engineering. Contractually specifying a business service is crucial for the design and operationalization of business services from the service providers' point of view. In this article, we present an overarching modeling and formalization approach to the contractual specifications of business services. First, business services are conceptually described from three different perspectives, giving rise to a list of service descriptors that matter most for the contractual specifications of services. Second, we formalize the service descriptors. Third, we devise a formal machinery to (a) verify if a group of services contractually match the specification of the bulkier service in question; (b) to assess the contractual proximity of service groups relative to a contractual service specification to help decide which combination of services from a catalog best realize the desired functionality.

Keywords: Service Engineering · Goal modeling · Service contract · Quality of service · Service composition · Serviceability

1 Introduction

In the last few years, service-oriented computing has become an emerging research topic in response to the shift from product-oriented economy to service-oriented economy. On the one hand, we now live in a growing services-based economy in which every product today has virtually a service component to it

© Springer-Verlag GmbH Germany 2017
A. Hameurlain et al. (Eds.): TLDKS XXXI, LNCS 10140, pp. 94–123, 2017.
DOI: 10.1007/978-3-662-54173-9_5

[40]. In this context, services are increasingly provided in different ways in order to meet growing customer demands. Business domains involving large and complex collection of loosely coupled services provided by autonomous enterprises are becoming increasingly prevalent [22,44]. On the other hand, Information Technology (IT) has now been thoroughly integrated into our daily life [17] and gradually gives rise to the paradigm of ubiquitous computing. As such, business services are essentially IT-enabled making the border between business services[1] and IT-enabled services blurred. At the high-level operationalization of a business service, we see business activities happening between service stakeholders. We may or may not witness IT operations at this representational level. At lower levels, the operationalization of these services are eventually translated into IT operations as we have seen in the cases of banking services, recruitment services, library services, auctioning services, etc.

From an IT perspective, there is a proliferation of methods and languages for Web services. Unfortunately, there has not been much work in modeling high-level services from a business perspective. The operationalization of business services has distinctive features that are not typically observed in plain Web services. Most notably, business services occur for a noticeable period of time, not spontaneously as Web services do. Their occurrences feature incremental human-mediated developments. As such, the representation of business services requires that we view human activity and human-mediated functionality through the lens of computing and systems engineering. Business services are typically operationalized by means of outsourcing or subcontracting, through which the provider of a relatively complex business service breaks it down into constituent services and subcontracts some of them to other service providers. Our study of real-life business services and collaboration with our industry partners reveal that, alternatively, the provider may pick up an existing set of services (e.g. from a service catalog) that best match it. The challenges here are to (i) verify if a group of services contractually match the specification of the big service in question; (ii) determine the preference in choosing a set of services from a service catalog in order to operationalize the big service in question.

This article substantially extends our previous work on the contractual proximity of business services [29]. In this paper, we first present a 3-dimensional modeling space that takes into account multiple perspectives of business services [30] – the basis of our line of research on business services. We then narrow down the scope of our research to the decomposition of services where provide both type-based reasoning and formality of service contracts including qualities of service and penalties. We devise a formal machinery to enable that verification and assessment as a continuation of our previous work on the representation of business services [14,31,32].

---

[1] By calling them business services, we mean services happening between people or business entities. They are enabled by IT in one way or another. For the sake of simplicity, we shall use the term "business service" or simply "service" to refer to these IT-enabled business services throughout this paper.

Paper Structure. Section 2 describes a running example and helps formulate our research questions. In Sect. 3, we give the preliminaries for the modeling and formalization of service contractual specifications. Section 4 reinforces the necessity of modeling high-level business services from multiple perspectives. In Sect. 5, we make explicit the context of our work where we reason about the decomposition of contractual specifications. Section 6 furthers our work by proposing a proximity scale to compare two groups of contractual specifications with respect to a certain service. We survey related work in Sect. 7. Section 8 ends the article by drawing some concluding remarks and outlining our future work.

2   From Service Modeling to Serviceability

In this section, we give the scope of our research – service design by means of decomposition taken from the service provider's point of view. We formulate our research questions in the defined research scope (Subsect. 2.2). Subsection 2.1 presents a running example that helps explain the formulated research problems. The example will be used throughout the article for further explanation of our framework.

2.1   Running Example

Let us consider a (passenger) car rental as a service denoted as S_1 provided by a car rental company. For the sake of service design and operationalization, the provider of this service may break it down into three constituent services: (s_{11}) identity check & deposit; (s_{12}) vehicle pickup & return; (s_{13}) vehicle maintenance. We reason about the contractual specifications of these high-level business services according to the multi-dimensional representational space presented in Subsect. 4. We however will not consider business processes that capture the operational details (e.g. order in which these constituent services may be operationalized).

Figure 1 illustrates our example. Suppose that all of these services have been contractually specified in terms of service descriptors ranging from propositional statements (e.g., service goals, pre- & post-conditions, assumption, penalty), service objects (e.g. input, output), quantifiable measurements (e.g. quality of service). The question is, given the contractual specifications of $\{s_{11}, s_{12}, s_{13}\}$, is the decomposition $S_1 = \{s_{11}, s_{12}, s_{13}\}$ valid? In other words, is the contractual specification of S_1 is serviceable by s_{11}, s_{12} and s_{13}? We will analyze this question on each service descriptors defined in Table 2.

2.2   Research Problems

Given a relatively complex business services, practitioners can deal with its complexity either by (a) breaking it down into constituent services through common practices such as outsourcing or delegation; (b) or picking up a set of services from an existing service catalog that best match it. In either way, we need to

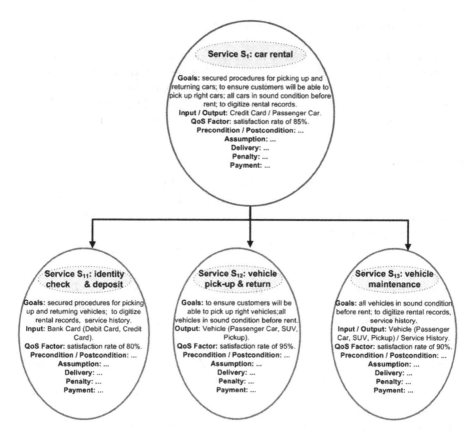

Fig. 1. Car rental service is broken down into component services: (s_{11}) identity check & deposit, (s_{12}) vehicle pickup & return and (s_{13}) vehicle maintenance.

justify whether a set of services contractually match or almost match the big service in question. We were first motivated by problem (a). In this paper, we devise a formal machinery to assess verify whether a set of constituent services match the contractual service specification of a business service in question. This work also sheds light on a potential solution to problem (b): how to assess the proximity of a set of services relative to a contractual service specification. We elaborate this standpoint as follows.

- Given a set of service models, is a given contract *serviceable*? We define *serviceability* as follows: a contract is serviceable given a set of services if and only the set of services represent a valid decomposition of the contract (which itself is viewed as a service)
- If contract is not serviceable, what is a minimally different set of service models that will make it serviceable?
- We represent contractual specifications using service descriptors that are different in nature. It is thus challenging to devise a single formal framework for reasoning about the decomposition of contractual specifications uniformly across all service descriptors.

3   Background

To verify whether a set of constituent services contractually meet the main service, we need to reason about the collective behavior of these services as well as their contractual violations. This sort of reasoning takes their root in the notion of behavioral subtyping, contracts monitoring and the semantics of business processes. We present the background for these topics in this section.

3.1   Subtyping

The notion of substitutability was made popular in object-oriented programming and later extended to the context of object-oriented conceptual modeling [8]. An object can substitute for another if the former can be safely used in a context where the latter is expected. This proposition is formulated in Definition 1. As an example, a passenger car can substitute for an vehicle. The type that describes all vehicles subsumes the type describing passenger cars.

Definition 1. *Object* s *can substitute for object* q, *denoted as Type(s) <: Type(q), if object* s *belongs to a subtype of or the same type as what object* q *belongs to. The type of* q *is said to subsume the type of* s.

Behavioral subtyping is about the behavioral substitutability of objects concerning a subtype and a supertype. An object of a subtype must behave compatibly to what it does if it is regarded as an object of the supertype. The notions of pre/post condition and invariant has widely been investigated in the subjects of actions modeling [23], design by contract [39] and Liskov and Wing's work on behavioral subtyping [37]. The subtyping relation can be defined as follows [37]

- Invariants of the supertype are respected in the subtype
- Subtype methods preserve the supertype methods' behavior. Syntactically, we have the contravariance rule on method arguments, the covariance rule on method result and the containment of exceptions. Semantically, the method's precondition is weakened in the subtype but the method's postcondition is strengthened.
- Subtype constraints entail supertype constraints

In business modeling, the concept of service assumption might sounds similar to the concept of invariant but they differ in nature. Like invariants, assumptions need to be monitored during the occurrence of a service. Unlike invariants (which must be respected during the service occurrence), if a violation of assumptions is detected, the service may have to be aborted or continue its course under different terms (e.g. going into exceptions).

3.2   Type and Prototype in Conceptual Modeling

> "...Essentially, all models are wrong, but some are useful..."
>
> – George E.P. Box

Type/instance modeling is a mechanism to group and describe objects having common properties. This way of reasoning has widely been adopted in the object-oriented paradigm, perhaps based on an assumption that classification is effective for conceptualizing business domains. This standpoint takes its root in philosopher Aristotle's work that aimed to categorize all natural things into a comprehensive taxonomy [46]. When it comes to engineering, the standpoint has another advantage – the computational efficiency of class inheritance in enabling the programmers to extend their programming classes.

In the common sense, a type defines properties that could be shared between instances. A class refers to the set of all instances that can be created out of a type. People usually make an implicit assumption in reasoning about type and class that all instances are treated equal.

There exists a different vision on this matter. As Taivalsaari stressed, the so-called Aristotelian theory usually suffices to deliver a conceptually elegant model for many domains [46]. However, classification might not work well for graded categories where one cannot state that an object fully meets all defining predicates of a category. Instead, as pointed out by Rosch, we say that a category consists of concrete objects having unequal status [42]. The object that best characterizes a category constitutes a prototype. When it comes to engineering, we may need a radically departed programming philosophy rather than the traditional object-oriented paradigm (e.g. cloning objects in place of class inheritance in Beta and Self) at the expense of computational efficiency.

Advocates of prototype-based modeling such as Lieberman [35] and Rosch [42] argue that psychologically, people tend to develop their conceptualization of a new subject by linking it to the closest examples they can think of. As such, the so-called type-based modeling seems to be artificial.

Consensus on these two theories is that the prototype theory might reflect real-world phenomena and business situations better than the Aristotelian theory does. However, the former limits the ability to communicate conceptual models among people (e.g. classes and generalization/specialization are widely used in meta-modeling). Also, the former is technologically expensive and generally unfamiliar among programmers.

3.3 Semantically Annotated Business Processes

An semantically annotated business process model is a process in which every task has been annotated with immediate effects. To determine the functionality delivered up to a given point of time during the occurrence of an annotated process, we reason about the cumulative effect. We suppose that analysts can associate context-independent effect to each step represented in the process. There exists a technique that contextualizes these effects, i.e., to compute cumulative effects. The technique, called ProcessSEER, involves doing two stages of computation [19]. In the first stage, we derive a set of possible *scenario label*(s) for the given point in the process view. Each scenario label is a precise lists of steps that define a path leading from the start point to a the point being considered. In the second stage, the contiguous sequence of steps in each scenario label is taken into account to accumulate effects annotated to steps along this scenario in a pair-wise fashion.

A functionality annotation states the effect of having functionality delivered at a specified task. The effect can be textual. Alternatively, it could be written in first-order logic (FOL) or some computer-interpretable form. The total functionality delivered up to a certain task is the accumulation of all effects of the precedent tasks. We assume that the delivery annotations have been represented in conjunctive normal form (CNF) where each clause is also a prime implicate (this provides a non-redundant canonical form) [41]. The cumulative effect of tasks can inductively be defined as follows. The cumulative effect of the very first task is equal to its delivery annotation. Let $\langle Tk_i, Tk_j \rangle$ be an ordered pair of consecutive tasks such that Tk_i precedes Tk_j; let e_i be an effect scenario associated with Tk_i and e_j be the delivery annotation associated with Tk_j. Without loss of generality, we assume that e_i and e_j are sets of clauses. The resulting cumulative effect, denoted by $acc(e_i, e_j)$ is defined as follows.

- $acc(e_i, e_j) = e_i \cup e_j$ if $e_i \cup e_j$ is logically consistent
- Otherwise $acc(e_i, e_j) = e_i' \cup e_j$ whereby $e_i' \subseteq e_i$ such that $e_i' \cup e_j$ $e_i' \subset e_i'' \subseteq e_i$ $e_i'' \subseteq e_i' \subseteq e_i$ such that $e_i'' \cup e_j$ is consistent

The task of accumulating functionality annotations is non-trivial since there might be various paths that can be traversed during the occurrence of the process up to the point of time being considered. We call the path leading to a certain point of time a *scenario label*. A scenario label can either be a sequence, denoted by the $\langle \rangle$ delimiters, or a set denoted by the $\{\}$ delimiters or combinations of both. The set delimiters are used to deal with parallel splits, and distinct elements in a set can be performed in any order [19]. Elements in a scenario label could be tasks (which have delivery annotations) or control elements (e.g. mutually exclusive split, parallel split). In addition to pair-wise effect accumulation across scenario labels, we need to make special provision for the following: (i) accumulation across AND-joins, and (ii) accumulation of effects over input/output flows.

3.4   Semiring

Semiring is a mathematical structure that features a domain plus two operations satisfying certain properties, as described in Definition 2.

Definition 2. *A semiring is a tuple* $\langle A, \oplus, \otimes, \bar{0}, \bar{1} \rangle$ *such that*

- *A is a set and $\bar{0}, \bar{1} \in A$*
- *$\oplus$, called the additive operation, is a commutative, associative operation having $\bar{0}$ as its neutral element (i.e. $a \oplus \bar{0} = a = \bar{0} \oplus a$)*
- *$\otimes$, called the multiplicative operation, is an associative operation such that 1 is its unit element and $\bar{0}$ is its absorbing element (i.e. $a \otimes \bar{0} = \bar{0} = \bar{0} \otimes a$)*
- *$\otimes$ distributes over $\oplus$ (i.e. $\forall a, b, c \in A \to a \otimes (b \oplus c) = a \otimes b \oplus a \otimes c$)*

An idempotent semiring is a semiring whose additive operation is idempotent (i.e. $a \oplus a = a$). This idempotence property allows us to endow a semiring with a canonical order defined as $a \preceq b$ iff $a \oplus b = b$ [10]. There exists another form of idempotent semiring called c-semiring whereby the $\oplus$ operator is defined over subsets of a domain and as such it has flattening property [4]. The endowed order of a c-semiring is actually a partial order that would be used for choosing "best" solutions in a constraint satisfaction problem.

Definition 3. *A semiring will be called c-semiring, where "c" stands for "constraint", meaning that they are the natural structures to be used when handling constraints [5]. A c-semiring is a tuple $\langle A, \oplus, \otimes, \bar{0}, \bar{1} \rangle$ such that*

- *A is a set and $\bar{0}, \bar{1} \in A$*
- *$\oplus$ is defined over (possibly infinite) sets of elements of A as follows:*
    - *For all $a \in A, \sum(\{a\}) = a$;*
    - *$\sum(\emptyset) = \bar{0}$ and $\sum(A) = \bar{1}$;*
    - *$\sum(\bigcup A_i, i \in I) = \sum(\{\sum(A_i), i \in I\})$ for all sets of indices I (flattening property).*
- *$\otimes$ is a binary, associative and commutative operation such that $\bar{1}$ is its unit element and $\bar{0}$ is its absorbing element*
- *$\otimes$ distributes over $\oplus$ (i.e. for any $a \in A$ and $B \subseteq A$, $a \otimes \sum(B) = \sum(\{a \otimes b, b \in B\})$)*

To make an idempotent semiring applicable for the representation of QoS, we endow it with a canonical order defined as $a \preceq b$ iff $a \oplus b = b$. A semiring is used to express the domain and the order between values that feature a QoS. To represent QoS factors, we may use the notion of *bounded lattice*. Each bounded lattice has a greatest element (denoted as $\top$) and a least element (denoted as $\bot$) and features two operations: meet (denoted as $\wedge$) and join (denoted as $\vee$) [10].

3.5   Business Contract Modeling Based on Deontic Logic

Business contracts specify obligations, permissions and prohibitions as mutual agreements between business parties [36], as well as actions to be taken when a contract is violated. Governatori et al. [15] have proposed such a contract modeling language which includes a non-boolean connective, $\odot$, to represent contrary-to-duty obligations (i.e., what should be done if the terms of a contract are violated). Deontic operators capture the contractual modality (i.e. obligations, permissions and prohibitions) [13]. Governatori et al. represent a contractual rule as $r : A_1, A_2 \ldots A_n \vdash C$ where each A_i is an antecedent of the rule and C is the consequent. Each A_i and C may contain deontic operators but connectives can only appear in C.

As an example, $r : \neg p, q \vdash O_{seller}\alpha \odot O_{seller}\beta$ is a contractual rule (identified by r) stating that if antecedents $\neg p$ and q hold, then a seller is obliged to make sure that α is brought about. Failure to do so results in a violation, for which a reparation can be made by bringing about β (the connective $\odot$ can therefore be informally read as "failing which").

Definition 4 *[15]. Contractual rules r and r' can be merged into rule r'' as follows where X denotes either an obligation or a permission.*

$$\frac{r : \Gamma \vdash O_s A \odot (\bigodot_{i=1}^{n} O_s B_i) \odot O_s C \quad r' : \Delta, \neg B_1, \neg B_2, \ldots, \neg B_n \vdash X_s D}{r'' : \Gamma, \Delta \vdash O_s A \odot (\bigodot_{i=1}^{n} O_s B_i) \odot X_s D}$$

The $\otimes$ operator is associative but not commutative. This property matters when reasoning about the subsumption and merging of contractual rules. Definition 4 defines how contract rules might be merged. Governatori et al. also devise a machinery for determining if one contractual rule subsumes another as presented in Definition 5.

Definition 5 *[15]. Let's consider two rules $r_1 : \Gamma \vdash A \odot B \odot C$ and $r_2 : \Delta \vdash D$ where $A = \bigodot_{i=1}^{m} A_i$, $B = \bigodot_{i=1}^{n} B_i$ and $C = \bigodot_{i=1}^{p} C_i$. Then r_1 subsumes r_2 (i.e. r_2 can safely be discarded if we have r_1) iff*

1. $\Gamma = \Delta$ and $D = A$; or
2. $\Gamma \cup \{\neg A_1, \ldots, \neg A_m\} = \Delta$ and $D = B$; or
3. $\Gamma \cup \{\neg B_1, \ldots, \neg B_n\} = \Delta$ and $D = A \odot \bigodot_{i=0}^{k \leq p} C_i$

4   On Business Services Modeling: The $3 \times 3 \times 3$ Grid

To make the paper self-contained, we present a multi-perspective representational grid that we base our work on to reason about service relationships [28]. This grid makes explicit that the occurrence of business services usually involves multiple stakeholders and sociotechnical systems. Primary stakeholders of a business service are service providers and service consumers. The providers and the consumers have different perspectives on a service they provide and consume, respectively. These two perspectives are the main sources of representational building blocks for business services. The context where services are provided/consumed is another perspective that addresses social issues of the business services being represented. In short, the provider, the consumer and the context of services form a 3-dimensional representational space of business services (see Fig. 2).

The dimension of the service consumers has three concepts: received value, service input/output and service bundling/unbundling. Service values capture the values added to consumers' business after they consumes a service. Services values are not tangible but they can be cognitively experienced or sensed and thus are rather subjective. Service input and output are items that are created or exchanged during the occurrence of a business service. Unlike service values, service input and output are tangible or perceivable. Business services can be consolidated into service bundles [26] – a practice that is called service bundling. Service unbundling refers to the practice of splitting a relatively big (and costly) service to give service consumers more choices in consuming certain parts of the service they really want to without unnecessarily making the full payment.

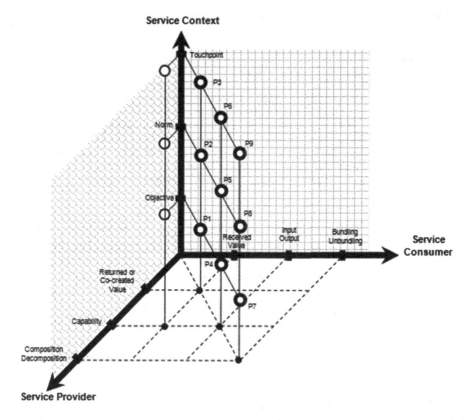

Fig. 2. The representational space of business services has a total of 27 representational points $(3 \times 3 \times 3)$.

In the dimension of the service providers, the concepts that matter are the returned/co-created value, the service capability and service composition/decomposition. Returned/co-created values of a business service capture what the service provider gains after providing the service. They could be returned values such as payments/acknowledgments or values that are co-created by the consumers of the service. Service capability is the modeling concept that permits reasoning about what a service does in order to deliver its values and provision its outputs. The concept of service capability takes its root from business modeling. Homann conceives the business as a network of capabilities [47]. Pohle stresses that business components have capabilities [12]. Service capability may not give imperative details of how service values and service outputs can be achieved. To formally describe service capability, we can annotate precondition (i.e. conditions in which the service may occur), postcondition (i.e. conditions and/or effects when the service has occurred) and schedules (i.e. checkpoints of functionality delivery, payments and penalties in case expected functionality is not delivered). On one hand, service decomposition refers to the practice of

Table 1. Intuitive meanings of the main nine points in the $3 \times 3 \times 3$ representational space.

Point	Explanation
P1	Business objectives relate values, co-created values and returned values of business services.
P2	Norms are assigned to co-created values and returned values of business services.
P3	Touchpoints help judge values, co-created values and returned values of business services.
P4	Business objectives help identify capabilities and input/output of business services.
P5	Norms can be put on capabilities and input/output of business services.
P6	Touchpoints help us define capabilities and input/output of business services.
P7	Business objectives give hint on how business services are (un-)bundled and (de-)composed.
P8	Norms help us judge the ways services are (un-)bundled and (de-)composed.
P9	Touchpoints help us identify potential (un-)bundling and (de-)composition of services.

breaking down a relatively big service with high complexity to facilitate the service design and service operationalization in a top-down fashion. On the other hand, service composition means putting relatively small services together to create a more complex service. The decomposition & composition may look similar to unbundling & bundling as both could be explained using the top-down & bottom-up principles. They matter on different perspectives.

The dimension of service context features objective, norm and touchpoint. We can bear on the analysis of why business services are designed, operationalized and consumed by adding business objectives to the representation of services. Service stakeholders (i.e. providers and consumers) interact with one another in order to create values and exchange input/output in a business service. Their behaviors in providing or consuming a service are regulated through norms. Obligations, prohibitions and permissions are typical norms. Another kind of norms that we find relevant in the dimension of service context is the concept of assumption. In our industry engagements with government agencies, we have found what were documented as "client responsibilities" in their service catalog could best be expressed as service assumptions. These assumptions state what service consumers are expected to do, in order to enable the service provider to fulfill relevant service capabilities. In general, norms need to be monitored during the occurrence of a business service - if a violation is detected, the service may have to be aborted. Service touchpoint is the place where service interactions happen [6]. Through touchpoints, the service is experienced and perceived with all the senses.

In our approach, a service representation can intuitively be interpreted as the set of points in the 3-dimensional space whose dimensions correspond to the service provider, the service consumer and the service context. There are a total of 9 points[2] in this representational grid enumerated as P_1, P_2... P_9 (each of them is depicted by small bold circles in the figure). In Table 1, we give intuitive meanings for these points.

---

[2] Theoretically, we could have up 27 points in this space. The nine points listed here make the most significant meaning when combining the three modeling perspectives.

This line of our research has resulted in a family of modeling languages for business services. On one hand, we proposed a representation language for formally reasoning about contractual specifications of business services [14]. On the other hand, we defined an informal service description language with an emphasis on service capabilities and properties [31]. In this work, we narrow down to a list of service descriptors in the contract-oriented representation of business services. These service descriptors serve as the basis for conceptually modeling and formally reasoning about the decomposition of business services. Table 2 describes them in detail. Each service descriptor has an informal definition. A service descriptor also features a set of perspective concepts taken from the $3 \times 3 \times 3$ representational grid that matter most.

5 Conceptual Modeling and Formalization for Contractual Specifications of Services

We devise a formal machinery to verify if a service decomposition is valid as a solution to problem (a) mentioned in Sect. 2.2. Intuitively, the combination of constituent services can substitute for the decomposed one without affecting the existing business that the service consumers engage in. In other words, the constituent services altogether deliver at least the same while they expect no more than what the decomposed service does. We propose conceptual modeling and formalization for reasoning about contractual specifications of business services with respect to the service descriptors listed in Table 2. The running example described in Subsect. 2.1 will be used for illustration purposes.

5.1 Goal

Following [34], assuming that a set of goals refers to the conjunction of its elements, we define goal refinement in Definition 6.

Definition 6. *Goal G is refined into a set of sub-goals $\{g_1, g_2, \ldots, g_n\}$ to be valid if and only if:*

- $g_1 \wedge g_2 \wedge \ldots \wedge g_n \not\models \perp$
- $g_1 \wedge g_2 \wedge \ldots \wedge g_n \models G$
- $G' \not\models G$ *for any* $G' \subset \{g_1, g_2, \ldots, g_n\}$

Example 1. Table 3 lists goals of service `car rental` in its leftmost two columns. Corresponding goals of services `identity check & deposit`, `vehicle pickup & return` and `vehicle maintenance` are given in the rightmost two columns. Note that all goals are described both informally (in natural language) and formally (by means of first-order logic). Given $Car \subset Vehicle$ and a few deduction rules in first-order logic [45], we can straightforwardly prove that the constituent services' goals actually refine `car rental`'s goals according to Definition 6.

Table 2. Service descriptors of a contract-oriented language specifically defined for the representation of business services.

Descriptor	Informal definition	Perspective concepts
Goal	Intended effects or achievements of the service being represented	Objective, Decomposition, Composition
Precondition	Conditions that must hold to enable the occurrence of the service being represented	Objective, Decomposition, Composition
Postcondition	Effects or achievements of a service. They must hold upon the completion of the service being represented.	Objective, Decomposition, Composition
Assumption	Conditions on whose validity the occurrence of the service is contingent, but whose validity might not be verifiable when the service is invoked or during its execution	Norm, Decomposition, Composition
Input	Tangible or perceivable items that are usually fed by service consumers during the occurrence of the service being represented	Input/Output, Decomposition, Composition
Output	Tangible or perceivable items that are created or exchanged during the occurrence of the service being represented	Input/Output, Decomposition, Composition
QoS factor	Non-functional properties of the service being represented. Each Quality-of-Service (QoS) factor can be described in terms of the upper and lower bounds with quantitative evaluations.	Touchpoint, Decomposition, Composition, Bundling, Unbundling
Delivery	Incremental functionality during the occurrence of the service being represented.	Norm, Decomposition, Composition
Payment	Incremental payment during the occurrence of the service being represented. It may be described as a schedule in the form $\langle amount, deadline \rangle$.	Touchpoint, Bundling, Unbundling
Penalty	A penalty is given when a functionality is not delivered as scheduled. It is usually associated with a delivery schedule.	Norm, Decomposition, Composition, Bundling, Unbundling

5.2   Precondition/Postcondition

The constituent services altogether require same or weaker preconditions while producing same or stronger postconditions as the decomposed service does. This proposition could formally be interpreted as follows.

Table 3. Informal and formal representation of service goals. Note that s_{11}, s_{12} and s_{13} refer to services `identity check & deposit`, `vehicle pickup & return` and `vehicle maintenance`, respectively.

Car rental's goals		Constituent services' goals		
To provide customers with secured procedures for picking up and returning rental cars	$\forall r \in Customer, \forall c \in Car : booked(r, c) \rightarrow creditcardDeposited(c)$	$\forall r \in Customer, \forall v \in Vehicle : booked(r, v) \rightarrow (creditcardDeposited(v) \vee bankDeposited(v))$	To provide customers with convenient yet secured procedures for picking up and returning rental vehicles	s_{11}
To ensure customers who booked in advance will be able to pick up cars they selected	$\forall r \in Customer, c \in Car : booked(r, c) \rightarrow proceedWithPickup(r, c)$	$\forall r \in Customer, v \in Vehicle : booked(r, v) \rightarrow proceedWithPickup(r, v)$	To ensure customers who booked in advance will be able to pick up vehicles they selected	s_{12}
To make sure that all cars are mechanically sound before handing them over tenants	$\forall r \in Customer, c \in Car : booked(r, c) \rightarrow mecSound(c)$	$\forall r \in Customer, v \in Vehicle : booked(r, v) \rightarrow mecSound(v)$	To make sure that all vehicles are mechanically sound before handing them over tenants	s_{12}, s_{13}
To computerize the pickup and return procedures and to digitize rental records	$\forall c \in Car : rentalDigitized(c)$	$\forall v \in Vehicle : rentalDigitized(v) \wedge historyDigitized(v)$	To computerize the pickup and return procedures and to digitize rental records as well as service history of all vehicles	s_{11}, s_{13}

- The pre-condition of the decomposed service entails those of its constituent services, formally $pre_S \models pre_1 \wedge pre_2 \wedge \ldots \wedge pre_n$ where $pre_1, pre_2, \ldots, pre_n$ denote preconditions of constituent services; pre_S denotes the precondition of the decomposed service.
- The post-conditions of constituent services entail that of the decomposed service, formally $post_1 \wedge post_2 \wedge \ldots \wedge post_n \models post_S$ where $post_1, post_2, \ldots, post_n$ denote postconditions of constituent services; $post_S$ denotes the postcondition of the decomposed service.

5.3   Assumption

The standpoint on service assumption is similar to that on service precondition. Intuitively, the constituent services altogether should make same or weaker as the decomposed service does. This proposition could be formalized as follows. The assumption of the decomposed service entails those of its constituent services, formally $asmp_S \models asmp_1 \wedge asmp_2 \wedge \ldots \wedge asmp_n$ where $asmp_1, asmp_2, \ldots, asmp_n$ denote assumptions made for constituent services; $asmp_S$ denotes the assumption made for the decomposed service.

Example 2. One of the assumptions of service `car rental` is about the customer's responsibility: tenants must check (and top up if necessary) engine oil and other fluids of their rental car especially during long drives and will be held liable for any breakdown caused by the insufficiency of oil and/or fluid. If the constituent service `vehicle maintenance` ensures that all vehicles are equipped with appropriate level sensors, it can bear on the following weaker assumption without affecting the serviceability of the contractual spec of service `car rental`: tenants must top up engine oil and other fluids of their rental car whenever oil/fluid level warnings are given on their car dashboard.

5.4   Input/Output

Conceptual modeling is about constructing representations of a subject for a particular purpose, e.g., designing a software system, capturing the enterprise context of an information system, showing a database schema. Typically, conceptual modeling is done for a system that is central to engineering and/or development activities – the system of interest. Modeling has a few purposes. Documentation is one of them, reasoning about the information captured is another and facilitating our understanding (and thus sharing knowledge) is yet another purpose. Modeling could be perceived as mapping from the subject, e.g. system of interest, to a representation done in a particular modeling language [16]. Whether this representation is precise and whether the mapping is semantics-preserving are open for further investigation.

Meta-modeling gained momentum as soon as we research on conceptual modeling [2,27]. In order to improve the readiness of enterprise models, meta-modeling has been extensively used in the representation and analysis of enterprise architecture [7,43].

The constituent services altogether require same or less input while produce same or more output as the decomposed service does. The intuitive meaning of "less input" can be explained as (i) constituent services take a subset of input objects taken by the decomposed service; or (ii) constituent services take the same number of input objects but some of their input objects subsume input objects of the decomposed services. Similarly, "more output" actually means (i) constituent services produce a superset of output objects produced by the decomposed service; or (ii) constituent services produce the same number of output objects but some of their output objects can be substituted for output objects of the decomposed services.

The notion of substitutability was made popular in object-oriented programming and later extended to the context of object-oriented conceptual modeling [37]. An object can substitute for another if the former can be safely used in a context where the latter is expected. This proposition is formulated in Definition 1. As an example, a passenger car can substitute for an vehicle. The type that describes all vehicles subsumes the type describing passenger cars.

Whether a set of constituent services match a decomposed service in terms of input/output can be formulated as follows. Example 3 illustrates this formulation.

- An input object taken by the constituent services is either passed by the decomposed service or substitutable for by a corresponding input object taken by the decomposed service. Formally, we have $\forall x \,:\, x \in \bigcup_{i=1}^{n} input_i \,\to\, Pass(S, x) \lor (\exists x' : x' \in input_S \land \neg Pass(S, x') \land Type(x') <: Type(x))$ where $Pass(serv, o)$ implies that service $serv$ produces object o and passes it as an input object to another service; $input_S$ denotes the set of input objects of the decomposed service and $input_i$ the set of input objects of the ith constituent service.
- For every output object produced by the decomposed service, there is a corresponding output object produced by one of the constituent services that it can substitute for. Formally, we have $\forall x \,:\, x \in output_S \,\to\, \exists x' : x' \in \bigcup_{i=1}^{n} output_i \land Type(x) <: Type(x')$ where $output_S$ denotes the set of output objects of the decomposed service and $output_i$ the set of output objects of the ith constituent service. Note that some output objects produced by the constituent services may be consumed by the decomposed service.

Example 3. The constituent services s_{11}, s_{12}, s_{13} may accept input objects and produce the output objects of different types than the main service S_1 does. Specifically, s_{11} can accept not only credit card but also debit card (e.g. Maestro) whilst s_{12}, s_{13} can deal with not only passenger cars but also utility cars (e.g. 4×4, pickup). In addition, s_{13} produces more output objects (i.e. Service History) than S_1 does. This service decomposition is still valid in terms of input and output despite the difference of input/output types between the constituent services and the main one.

Figure 3 depicts the input and output objects of these services. We use dashed lines for the graphic of main service car rental and its input/output objects in

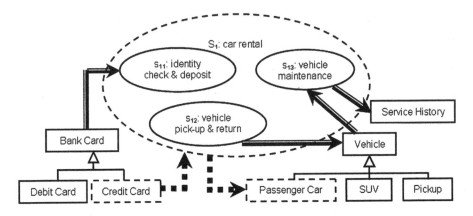

Fig. 3. The contractual spec of service `car rental` is serviceable by services `identity check & deposit`, `vehicle pickup & return` and `vehicle maintenance` in terms of input/output because the input and output objects of the constituent services subsume those specified for the decomposed service.

this figure. The constituent services and their input/output objects are drawn under solid lines. Double line arrows show input or output (depending on the direction of the arrows) of services. Triangle-headed arrows diagrammatically illustrate the subtyping relation between input/output objects.

5.5   QoS Factor

The QoS factors of the decomposed service must be satisfied by those of the constituent services. This proposition will be formally elaborated in light of the semiring-based representation of QoS factors. Definition 2 defines *semiring* [18] that will be used to express QoS factors.

Example 4. Service S_1 (`car rental`) and its constituent service s_{12} (`vehicle pickup & return`) has a QoS factor in common, which is about the likelihood that the customers will be handed a car of the specific type they have booked. S_1 aims to satisfy their customers' bookings at the rate of 85% or more while service s_{12} features a booking satisfaction rate of at least 95%. The domain of these two QoS factors is represented using an idempotent semiring denoted as $\langle [0,1], max, \times, 0, 1 \rangle$. This semiring takes the max function as its additive operation and the classical multiplication as its multiplicative operation. Intuitively, the max function of this semiring can be used to express a canonical order and the $\times$ operator is useful for reasoning over probability. Note that $[0,1]$ denotes an interval of real numbers that represents the probability domain. This semiring is endowed with the classical comparison as its canonical order.

    To this end, comparing the two QoS factors of these two services would boil down to checking if the lattice that represents the QoS factor of one service is a *sublattice* of the lattice that represents the corresponding QoS factor of the

Table 4. An example of how QoS factors can be expressed using semirings and lattices

Service QoS	Semiring $\langle A, \oplus, \otimes, 0, 1 \rangle$	Canonical order	Lattice $\langle \bot, \top, \wedge, \vee \rangle$
s_{12}: satisfaction rate of 95%	$\langle [0,1], max, \times, 0, 1 \rangle$	Real numbers comparison	$\langle 0.95, 1, min, max \rangle$
S_1: satisfaction rate of 85%	$\langle [0,1], max, \times, 0, 1 \rangle$	Real numbers comparison	$\langle 0.85, 1, min, max \rangle$

other service. As illustrated in Table 4, the QoS factor of s_{12} entails that of S_1 because the lattice created for s_{12} is in fact a sublattice of the lattice created for S_1.

Example 4 motivates us to define whether a set of constituent services fulfill the decomposed service in terms of QoS factors as follows. A set of constituent services fulfill the decomposed service in terms of QoS factors iff for each QoS factor of the decomposed service S, the following two clauses hold

- there is a corresponding QoS factor that is specified for the constituent service s_i of which the domain is represented using the same semiring
- the lattice that characterizes the QoS factor of S is a sublattice of the corresponding lattice for s_i

5.6   Delivery

We come up with a notion of *functionality entailment*. At any given time, the combined functionality delivered by the constituent services must entail the functionality specified for the decomposed service.

Let $df_i(t)$ denote the functionality delivered by the ith constituent service at the moment t. Let $df_S(t)$ be the functionality tentatively delivered at the moment t by the decomposed service. The functionality of a service as a whole can be regarded as a set of schedules each of which is given in the form $\langle functionality, deadline \rangle$. As such, we have $df_i(t) = \{\langle functionality_{ij}, deadline_{ij} \rangle | deadline_{ij} < t\}$, $df_S(t) = \{\langle functionality_{Sj}, deadline_{Sj} \rangle | deadline_{Sj} < t\}$.

To notion of functionality entailment can formally be captured as $\forall t \in Time :$ $acc(\{df_1(t), df_2(t), \ldots df_n(t)\}) \models df_S(t)$. Note that function acc yields cumulative effects of the functionalities of all constituent services (see Subsect. 3.3).

Example 5. Let us analyze the running example in the descriptor of service delivery. We analyze the functionalities delivered by services s_{11}, s_{12}, s_{13} and S_1, as depicted in Fig. 4. Table 5 explains the service delivery of these services in first order logic and the notion of functional entailment between them.

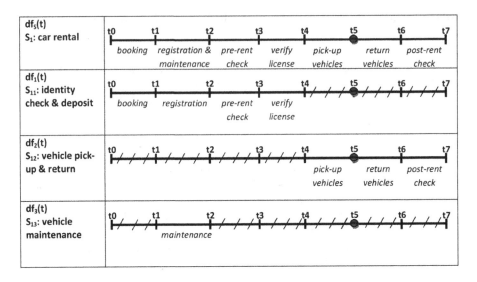

Fig. 4. Funtionalities of services s_{11}, s_{12}, s_{13} and S_1 delivered at different moments of time.

5.7   Penalty

Penalties will be applied if a service functionality is not delivered as scheduled. Penalties can be applied to constituent services as well as the decomposed

Table 5. Service delivery for the car rental example expressed in first order logic

$df_S(t_5)$	$\{<$ booking$, t_1$ $>,<$ registration&maintenance$, t_2 >$ $,<$ pre $-$ rentcheck$, t_3$ $>$ $,<$ verifylicense$, t_4$ $>,<$ pick $-$ upvehicles$, t_5 >\}$	$\exists r \in$ Customer$, c \in$ Car : booked$(r, c) \wedge$ (creditcardDeposited$(r) \vee$ bankDeposited$(r)) \wedge$ mecSound$(c) \wedge$ licensePresented$(r) \wedge$ rented(r, c)
$df_1(t_5)$	$\{<$ booking$, t_1$ $>,<$ registration$, t_2$ $>,<$ pre $-$ rentcheck$, t_3$ $>,<$ verifylicense$, t_4 >\}$	$\exists r \in$ Customer$, c \in$ Car : booked$(r, c) \wedge$ (creditcardDeposited$(r) \vee$ bankDeposited$(r)) \wedge$ mecSound$(c) \wedge$ licensePresented(r)
$df_2(t_5)$	$\{< pick - upvehicles, t_5 >\}$	$\exists r \in$ Customer$, c \in$ Car : rented(r, c)
$df_3(t_5)$	$\{< maintenance, t_2 >\}$	$\exists c \in$ Car : mecSound(c)
Therefore $acc(df_1(t_5), df_2(t_5), df_3(t_5))$	$\{<$ booking$, t_1$ $>,<$ registration$, t_2$ $>,<$ maintenance$, t_2$ $>,<$ pre $-$ rentcheck$, t_3$ $>$ $,<$ verifylicense$, t_4$ $>,<$ pick $-$ upvehicles$, t_5 >\}$	$\exists r \in$ Customer$, c \in$ Car : booked$(r, c) \wedge$ (creditcardDeposited$(r) \vee$ bankDeposited$(r)) \wedge$ mecSound$(c) \wedge$ licensePresented$(r) \wedge$ rented(r, c)

So we have $acc(df_1(t_5), df_2(t_5), df_3(t_5)) \models df_S(t_5)$

service. To represent service penalties, we use Deontic logic as presented in Subsect. 3.5. Each penalty is associated with a service delivery. They can together be represented as a contractual rule. Formally, $r_i(t) : \neg df_i(t) \vdash \bigodot_{j=1}^{n} O_i B_{ij}$.

Example 6. Let us consider the moment in the rental process when a customer is about to pick up a rental car. Services s_{11}, s_{12} and s_{13} are expected to deliver confirmed, picked and serviced respectively. The contractual rule for delivering pick of service s_{12} (see Fig. 1) can be formally expressed as

$$r_2 : \neg picked \vdash O_2 AltCar \odot O_2 Charged$$

Where $\neg picked \equiv \exists r \in Customer, c \in Car : booked(r,c) \wedge licensed(r) \wedge \neg rented(r,c)\}$ is as depicted in Fig. 4 and $AltCar \equiv \forall c \in Car, r \in Customer : booked(r,c) \wedge licensed(r) \wedge \neg rented(r,c) \rightarrow (\exists ac \in Car : comparableClass(ac,c) \wedge mecSound(ac) \wedge rented(r,ac))$.

This obligation rule states that a car of the class booked by the customer must be handed in, otherwise an alternative car of a comparable class (i.e. either the same class or a higher one than what was selected by the customer) shall be given to the customer. If this obligation is violated, the provider of s_{12} will be charged an amount of money.

The formal representation of the contractual rule for service s_{13} is as follows.

$$r_3 : \neg serviced, \neg AltCar \vdash O_3 QuickServ \odot O_3 Charged$$

Where $\neg serviced \equiv \exists r \in Customer, c \in Car : booked(r,c) \wedge \neg mecSound(c)$ and $QuickServ \equiv \forall r \in Customer, c \in Car : booked(r,c) \wedge \neg mecSound(c) \rightarrow quickServiced(c)$ (i.e. all cars, booked by the customers, that are not mechanically sound shall be serviced quickly). Given that service s_{11} does not have a contractual rule, merging r_2, r_3 will yield

$$r_{23} : \neg picked, \neg serviced \vdash O_2 AltCar \odot O_3 QuickServ \otimes O_3 Charged$$

according to Definition 4.

Now, let us formally represent the contractual rule for the main service in the following.

$$r_S : \neg delivery \vdash O_S AltCar \odot O_S DiscountNextRent$$

where $\neg delivery \equiv \exists r \in Customer, c \in Car : (booked(r,c) \wedge \neg mecSound(c)) \vee (booked(r,c) \wedge licensed(r) \wedge \neg rented(r,c)) \vee (booked(r,c) \wedge \neg mecSound(c) \wedge \neg quickServiced(c))$

We have $\neg QuickServ \equiv \exists r \in Customer, c \in Car : booked(r,c) \wedge \neg mecSound(c) \wedge \neg quickServiced(c)$. According to the definitions of $\neg picked$, $\neg serviced$, $\neg delivery$, $\neg QuickServ$ and function acc in Subsect. 3.3, we can prove $\{\neg serviced, \neg picked\} \cup \neg QuickServ = \neg delivery$. If the amount of money charged to s_{13} exceeds the mount of discount offered to the customer then according to criterion 3 of Definition 5, r_{23} subsumes r_S.

5.8   Payment

The service provider will ultimately take profit in outsourcing constituent services. In other words, the total payments the provider makes for outsourced services should be less than the payment received from the service consumers of the main service. The two monitoring approaches discussed in Subsect. 5.7 also apply to payments. In the linear approach, we have $\forall t \in Time : \sum_{i=1}^{n} pay(f_i(t)) > pay(f_S(t))$ where pay denotes the payment for a given functionality delivered. In the cumulative approach, the main service provider may accept a deficit in terms of payments during the occurrence of the service being decomposed. This can be formulated as $\sum_{i=1}^{n} acc(pay(f_i)) > acc(pay(f_S))$.

5.9   Formally Combining QoS, Payment and Penalty

In this subsection, following the rationale given in our previous work on the representation of contracts and goals [33], we present how to semantically represent the quality of service, payment and penalty of business services altogether using the mathematical structure of semiring. In order to facilitate the combination, we have aggregated the service descriptors QoS, Payment, Penalty as follows.

- The satisfaction has become an aspect of quality of service. It has been proven by a number of quality research services related to customer satisfaction: $Satisfaction \in [0, 1]$;
- Cost is a payment or price of service which is in the closure of cost values under addition. Let $C_0 = \{c_1, c_2, ...c_n\}$ be the initial set of values of Cost. By closure of C_0, we mean the smallest set containing all summation of elements in C_0: $C_0^+ = \{\sum_{k=1}^{\infty} (C_{i_1} + ... + C_{i_k}) | C_{i_k} \in C_0\}$. Based on the definitions, Cost is defined as $C = C_0^+ \cap [0, Cost_{max}]$, where $Cost_{max}$ is the highest cost that the customer can pay.
- The penalty clauses in contract are also represented as contractual rules. Rules are formulas in first order logic.

Mathematicaly combining Penalty, Payment and QoS (which are Rule, Cost and Satisfaction, respectively) results in the set of 3-tuple $A = \{\langle R, C, S \rangle\}$, where R is Rule, C is Cost, S is Satisfaction.

We consider the following ordering on set A. Let $a = \langle r_1, c_1, s_1 \rangle$ and $b = \langle r_2, c_2, s_2 \rangle$. We can say that $a \leq b$ iff $r_2 \vdash r_1$ or $(r_2 \nvdash r_1) \wedge (c_1 \leq c_2)$ or $(r_2 \nvdash r_1) \wedge (c_1 = c_2) \wedge (s_1 \leq s_2)$. In the case that $(r_2 \nvdash r_1) \wedge (r_1 \nvdash r_2)$, we define $a = b$.

Clearly, the relation "$\leq$" defines a total ordering over the set A. We define the $\oplus$ operation as the max operation with respect to this order.

The $\otimes$ operator is the multiplication acting on each component an element in the set A differently. The $\otimes$ operator's action on S is the min operation. The $\otimes$ operator's action on C is ordinary addition. The $\otimes$ operator's action on R is the merging of two different rules into one rule. More precisely, let $a = \langle r_1, c_1, s_1 \rangle$ and $b = \langle r_2, c_2, s_2 \rangle$, then $a \otimes b$ is defined as follow

$$a \otimes b := \langle \text{merge}(r_1, r_2), \ c_1 + c_2, \ \min\{s_1, s_2\} \rangle.$$

Proposition 1. *Let $\bar{0} = \langle \top, 0, 0 \rangle$, $\bar{1} = \langle \bot, 0, 1 \rangle$, with $\top$ and $\bot$ are tautology and the empty set, respectively. Then the tuple $\langle A, \oplus, \otimes, \bar{0}, \bar{1} \rangle$ is a semiring.*

Proof. Clearly, $\bar{0}$ is the neutral element of $\oplus$ and $\bar{1}$ is the unit element of $\otimes$. It suffices to show that the $\otimes$ operator is distributive over $\oplus$. Let $a = \langle r_1, c_1, s_1 \rangle$, $b = \langle r_2, c_2, s_2 \rangle$, and $c = \langle r_3, c_3, s_3 \rangle \in A$. We show that $a \otimes (b \oplus c) = (a \otimes b) \oplus (a \otimes c)$.

Indeed, without lost of generality, we may assume that $b \oplus c = c$ (i.e., $b \leq c$) and $b \neq c$. Then we have $r_3 \vdash r_2$. Therefore, $\text{merge}(r_1, r_3) \vdash \text{merge}(r_1, r_2)$. Hence

$$
\begin{aligned}
\text{R.H.S.} &= \max(\langle \text{merge}(r_1, r_2), \ c_1 + c_2, \ \min\{s_1, s_2\} \rangle, \\
&\qquad \langle \text{merge}(r_1, r_3), \ c_1 + c_3, \ \min\{s_1, s_3\} \rangle) \\
&= \langle \text{merge}(r_1, r_3), \ c_1 + c_3, \ \min\{s_1, s_3\} \rangle \\
&= a \otimes c \\
&= \text{L.H.S.}
\end{aligned}
$$

6 Towards Operationalization Preference and Contractual Proximity of Business Services

In this section, we address problem (b) that is articulated in the introduction. We define partial orders among service groups that fully meet a given contractual specification (Subsect. 6.1) and between those that do not (Subsect. 6.2). These definitions will potentially lead to selection criteria to assist the service provider in deciding which service group to be selected among those that are available from a service catalog in order to operationalize the given contractual specification.

6.1 Operationalization Preference

Suppose that we have multiple choices in picking up an existing set of services to match a given contractual service specification. In order to decide how to operationalize this contractual specification, the potential service provider for this contractual specification needs to know which service group is the best. We investigate the preference among groups of services that all satisfy a contractual specification. The preference is defined with respect to an individual service descriptor.

Table 6 gives formal definitions for this preference. A set of service SS_1 is preferred over another set SS_2 can be interpreted as if SS_1 satisfies an imaginary contractual specification represented by SS_2, in pretty much the same manner we define how SS_1 meets a contractual specification S in Sect. 5. Note that the preference between two sets of services with respect to input/output is defined based on Theorem 1 that is stated and proved as follows.

Theorem 1. *Substitutability as defined in Definition 1 is reflexive and transitive.*

Proof. For any object s, we obviously have $Type(s) <: Type(s)$. So s can substitute for itself. Suppose object s can substitute for object p, which can in turn substitute for object q. According to Definition 1, we have $Type(s) <: Type(p)$ and $Type(p) <: Type(q)$ meaning that any proposition that holds for object q's type will hold for object p's type and then for object s's type too. So $Type(s) <: Type(q)$ meaning s can substitute for q.

Each preference relationship defined in Table 6 features a partial order between sets of services in the sense that, between two sets of services that both satisfy a contractual specification, we may or may not firmly determine that one is preferred over the other.

Table 6. A set of services $SS_1 = \{s_{11}, s_{12}, \ldots, s_{1n}\}$ is preferred over another set $SS_2 = \{s_{21}, s_{22}, \ldots, s_{2m}\}$ that both satisfy a contractual service specification S.

Descriptor	Denotation	Definition
Goal	$SS_1 \succeq_{goal(S)} SS_2$	$goal_{11} \wedge goal_{12} \wedge \ldots \wedge goal_{1n} \models goal_{21} \wedge goal_{22} \wedge \ldots \wedge goal_{2m} \models goal_S$
Precondition	$SS_1 \succeq_{pre(S)} SS_2$	$pre_S \models pre_{21} \wedge pre_{22} \wedge \ldots \wedge pre_{2m} \models pre_{11} \wedge pre_{12} \wedge \ldots \wedge pre_{1n}$
Postcondition	$SS_1 \succeq_{post(S)} SS_2$	$post_{11} \wedge post_{12} \wedge \ldots \wedge post_{1n} \models post_{21} \wedge post_{22} \wedge \ldots \wedge post_{2m} \models post_S$
Assumption	$SS_1 \succeq_{asmp(S)} SS_2$	$asmp_S \models asmp_{21} \wedge asmp_{21} \wedge \ldots \wedge asmp_{2m} \models asmp_{11} \wedge asmp_{12} \wedge \ldots \wedge asmp_{1n}$
Input	$SS_1 \succeq_{input(S)} SS_2$	$\forall x : x \in \bigcup_{i=1}^{n} input_{1i} \rightarrow Pass(S, x) \vee (\exists x' : x' \in \bigcup_{j=1}^{m} input_{2j} \wedge \exists x'' : x'' \in input_S \wedge \neg Pass(S, x') \wedge Type(x'') <: Type(x') <: Type(x))$
Output	$SS_1 \succeq_{output(S)} SS_2$	$\forall x : x \in \bigcup_{j=1}^{m} output_{2j} \rightarrow \exists x' : x' \in \bigcup_{i=1}^{n} output_{1i} \wedge \exists x'' \in output_S \wedge Type(x) <: Type(x') <: Type(x'')$
QoS	$SS_1 \succeq_{qos(S)} SS_2$	for each QoS constraint of a member service $s_{1i} \in SS_1$, the following two clauses hold – there is a corresponding QoS constraint that is specified for a member service of $s_{2j} \in SS_2$ and another corresponding QoS constraint specified for S of which the domains are represented using the same semiring; – the lattice that characterizes the QoS constraint of s_{1i} is a sublattice of the corresponding lattice for s_{2j}, which is another sublattice of the corresponding lattice for S
Delivery	$SS_1 \succeq_{delivery(S)} SS_2$	$\forall t \in Time : acc(\{f_{11}(t), f_{12}(t), \ldots f_{1n}(t)\}) \models acc(\{f_{21}(t), f_{22}(t), \ldots f_{2m}(t)\}) \models f_S(t)$
Payment	$SS_1 \succeq_{payment(S)} SS_2$	$\forall t \in Time : \sum_{i=1}^{n} pay(f_{1i}(t)) > \sum_{j=1}^{m} pay(f_{2j}(t)) > pay(f_S(t))$
Penalty	$SS_1 \succeq_{penalty(S)} SS_2$	$\forall t \in Time : \sum_{i=1}^{n} pnlt(f_{1i}(t)) > \sum_{j=1}^{m} pnlt(f_{2j}(t)) > pnlt(f_S(t))$

6.2 Contractual Proximity

Given two sets of services that do not satisfy the contractual service specification S, we need to measure how far away they are different from S to determine which one is preferred over the other. We call this measure the *contractual proximity*.

Table 7 formally defines the contractual proximity of a set of business services with respect to a certain contractual service specification for individual service descriptors. The rationale behind this formal definitions is to determine smallest sets (for input and output), weakest conditions (for goal, precondition, postcondition and assumption) or minimal values (for payment and penalty) that make a set of services satisfy the given contractual service specification.

We now proceed in defining the preference between two sets of business services that do not satisfy a contractual service specification. Table 8 gives formal definitions of whether a set of services $CS_1 = \{s_{11}, s_{12}, \ldots, s_{1n}\}$ is preferred over another set of service $CS_2 = \{s_{21}, s_{22}, \ldots, s_{2m}\}$ with respect to a contractual service specification S. This preference is defined based on the definitions of contractual proximity given in Table 7. Intuitively speaking, between two sets of services that do not meet a contractual specification, the one that is closer to this specification will be preferred over the other.

7 Related Work

To the best of our understanding, service contracts have been studied in the context of Web service evolution [1] and the integration of heterogeneous services [9]. Our work differs from their largely of a view in which IT-enabled business services are contractually represented. We stress that human-mediated services should be contractually specified as quite differently as pure IT services (e.g. Web services) have been. Our research also differs from work on integrating human activities to business processes like BPEL4People [25] and human-mediated business processes [21]. We take a rather declarative approach, being in contrast with the imperative nature of the execution of business processes in these pieces of work.

In terms of QoS representation, semiring has been employed in expressing the QoS factors [11, 20]. In our approach, we tend to aggregate the service descriptors of QoS, payment and penalty and make them technically comparable. Additionally, QoS-based service discovery has been major research line in the community of Web services. Drivers for service selection vary from reputation [48], ontology [3] to monitoring [38]. Our framework could lead to an implementation of the discovery of a group of business services that are most suitable for a service contract, hence the notion of contractual proximity. Again, we take the concept of high-level, human-mediated business services as opposed to Web services in the mainstream research of service discovery. By aggregating a few service descriptors and provide a formal semantics for the ordering based on this aggregation, we propose sophisticated sorting machinery for handling a large number of combinations of services (or service bundles).

Table 7. Definition of contractual proximity of a set of services $CS = \{s_1, s_2, \ldots, s_n\}$ with respect to a contractual service specification S.

Descriptor	Denotation	Definition
Goal	$CS\Delta_{goal}S$	some goal gl s.t. — $gl \wedge goal_1 \wedge goal_2 \wedge \ldots \wedge goal_n \models goal_S$ — $\forall gl' : (gl' \wedge goal_1 \wedge goal_2 \wedge \ldots \wedge goal_n \models goal_S) \rightarrow (gl' \models gl)$
Precondition	$CS\Delta_{pre}S$	some condition $cond$ s.t. — $pre_S \wedge cond \models pre_1 \wedge pre_2 \wedge \ldots \wedge pre_n$ — $\forall cond' : (pre_S \wedge cond' \models pre_1 \wedge pre_2 \wedge \ldots \wedge pre_n) \rightarrow (cond' \models cond)$
Postcondition	$CS\Delta_{post}S$	some condition $cond$ s.t. — $cond \wedge post_1 \wedge post_2 \wedge \ldots \wedge post_n \models post_S$ — $\forall cond' : (cond' \wedge post_1 \wedge post_2 \wedge \ldots \wedge post_n \models post_S) \rightarrow (cond' \models cond)$
Assumption	$CS\Delta_{asmp}S$	some assumption Ap s.t. — $asmp_S \wedge Ap \models asmp_1 \wedge asmp_2 \wedge \ldots \wedge asmp_n$ — $\forall Ap' : (asmp_S \wedge Ap' \models asmp_1 \wedge asmp_2 \wedge \ldots \wedge asmp_n) \rightarrow (Ap' \models Ap)$
Input	$CS\Delta_{input}S$	a set of objects $In = \{x \mid x \in \bigcup_{i=1}^{n} input_i \wedge \neg Pass(S, x) \wedge \neg \exists x' \in input_S : Type(x') <: Type(x)\}$
Output	$CS\Delta_{output}S$	a set of objects $Out = \{x \mid x \in output_S \wedge \neg \exists x' \in \bigcup_{i=1}^{n} output_i : Type(x) <: Type(x')\}$
QoS	$CS\Delta_{qos}S$	a set of QoS constraints of any member service $s_i \in CS$ s.t. — there is no corresponding QoS constraint that is specified for S of which the domains can be represented using the same semiring; or — the lattice that characterizes the QoS constraint of s_i is not a sublattice of the corresponding lattice for S
Delivery	$CS\Delta_{delivery}S$	a function $dd(t)$ s.t. — $dd(t) \wedge acc(\{f_1(t), f_2(t), \ldots f_n(t)\}) \models f_S(t)$ — $\forall fct : (fct \wedge acc(\{f_1(t), f_2(t), \ldots f_n(t)\}) \models f_S(t)) \rightarrow (fct \models dd(t))$ where $t \in Time$
Payment	$CS\Delta_{payment}S$	a function $dpay(t) = \sum_{i=1}^{n} pay(f_i(t)) - pay(f_S(t))$ where $t \in Time$
Penalty	$CS\Delta_{penalty}S$	a function $dpnlt(t) = \sum_{i=1}^{n} pnlt(f_i(t)) - pnlt(f_S(t))$ where $t \in Time$

Table 8. A set of services $CS_1 = \{s_{11}, s_{12}, \ldots, s_{1n}\}$ is preferred over another $CS_2 = \{s_{21}, s_{22}, \ldots, s_{2m}\}$ with respect to a service specification S for individual service descriptors.

Descriptor	Denotation	Definition
Goal	$CS_1 \succeq_{goal} CS_2$	$CS_2 \Delta_{goal} S \models CS_1 \Delta_{goal} S$
Precondition	$CS_1 \succeq_{pre} CS_2$	$CS_2 \Delta_{pre} S \models CS_1 \Delta_{pre} S$
Postcondition	$CS_1 \succeq_{post} CS_2$	$CS_2 \Delta_{post} S \models CS_1 \Delta_{post} S$
Assumption	$CS_1 \succeq_{asmp} CS_2$	$CS_2 \Delta_{asmp} S \models CS_1 \Delta_{asmp} S$
Input	$CS_1 \succeq_{input} CS_2$	$CS_1 \Delta_{input} S \subseteq CS_2 \Delta_{input} S$
Output	$CS_1 \succeq_{output} CS_2$	$CS_1 \Delta_{output} S \subseteq CS_2 \Delta_{output} S$
QoS	$CS_1 \succeq_{qos} CS_2$	$CS_1 \Delta_{qos} S \subseteq CS_2 \Delta_{qos} S$
Delivery	$CS_1 \succeq_{delivery} CS_2$	$CS_2 \Delta_{delivery} S \models CS_1 \Delta_{delivery} S$
Payment	$CS_1 \succeq_{payment} CS_2$	$CS_1 \Delta_{payment} S \leq CS_2 \Delta_{payment} S$
Penalty	$CS_1 \succeq_{penalty} CS_2$	$CS_1 \Delta_{penalty} S \leq CS_2 \Delta_{penalty} S$

8 Conclusion

In our view, contractual concerns appear routinely in the description of business services, and are part of the discourse on service design and re-design. We have defined a family of representational languages dedicatedly for business services in a contract-oriented perspective [14, 31, 32]. Our work takes root in services science where we initially proposed a multi-perspective representational approach [30]. In this line of research, we are interested in the *serviceability* of contractual service specifications and motivated by the following two problems: (i) to formally reason about the decomposition of contractual specifications uniformly across all service descriptors; (ii) to verify a set of services against a certain contractual specification; (iii) to determine the preference in choosing a set of services from a service catalog in order to operationalize the contractual specification. In this paper, we present a formal machinery to verify the decomposition of services and assess the *contractual proximity* of business services against a contractual specification. The originality of our work is that we take into account the incremental nature of business services (i.e. delivery schedules, payment schedules, penalties) as well as human-mediated factors (i.e. QoS, assumption, goals) of business services in addition to service descriptors that have been commonly addressed in the field of service-oriented computing such as input/output, pre/post conditions.

Validation. Applications of our framework could be found in situation where a large number of business services are documented in a service catalog. In our previous work [28], we represent business services provided by an agency under the government body of an Australian state using the service descriptors defined in this article. Work are underway to extract textual description specifying the

QoS factors and the penalty rules from this service catalog for further reasoning[3] as formally presented in Subsect. 5.9.

In another line of our research, we look at the formal semantics of service bundling & unbundling. Work is currently underway to work out semirings that mathematically capture the quality of service, payment and penalty together in this regard. To make our framework computer-interpretable, we map our formal definitions to Alloy – a lightweight formal, declarative modeling language [24]. The goal is to develop a computer-interpretable engine that computerize our formal machinery. This will open the door for the implementation of a toolkit that manages a service repository and permits reasoning on service contracts (e.g. verifying a set of services against a contract, ranking alternative sets of services based on their serviceability on a specific contract). Another piece of future work is to support evolution of business services and of service contracts – changes made to any service of a service group that operationalize a contract shall be appropriately propagated to the contract and vice versa.

Acknowledgment. The first author would like to thank my former colleague, Aditya Ghose, for his valuable feedback on this work, especially in the formalization of service goals and goal entailment.

References

1. Andrikopoulos, V., Benbernou, S., Papazoglou, M.P.: Evolving services from a contractual perspective. In: Eck, P., Gordijn, J., Wieringa, R. (eds.) CAiSE 2009. LNCS, vol. 5565, pp. 290–304. Springer, Heidelberg (2009). doi:10.1007/978-3-642-02144-2_25
2. Atkinson, C., Kühne, T.: The essence of multilevel metamodeling. In: Gogolla, M., Kobryn, C. (eds.) UML 2001. LNCS, vol. 2185, pp. 19–33. Springer, Heidelberg (2001). doi:10.1007/3-540-45441-1_3
3. Bilgin, A.S., Singh, M.: A DAML-based repository for QoS-aware semantic web service selection. In: Proceedings of the IEEE International Conference on Web Services. Institute of Electrical and Electronics Engineers (IEEE) (2004)
4. Bistarelli, S., Montanari, U., Rossi, F.: Semiring-based constraint satisfaction and optimization. J. ACM **44**(2), 201–236 (1997)
5. Bistarelli, S.: Soft constraint satisfaction problems. In: Bistarelli, S. (ed.) Semirings for Soft Constraint Solving and Programming. LNCS, vol. 2962, 1st edn, pp. 21–50. Springer, Heidelberg (2004). Chap. 2
6. Bitner, M.: Evaluating service encounters: the effects of physical surroundings and employee responses. J. Mark. **54**(2), 69–82 (1990)
7. Buckl, S., Buschle, M., Johnson, P., Matthes, F., Schweda, C.: A Meta-language for enterprise architecture analysis. In: Halpin, T., Nurcan, S., Krogstie, J., Soffer, P., Proper, E., Schmidt, R., Bider, I. (eds.) Enterprise, Business-Process and Information Systems Modeling, vol. 81, pp. 511–525. Springer, New York (2011)
8. Cardelli, L., Wegner, P.: On understanding types, data abstraction, and polymorphism. ACM Comput. Surv. MIT Press Sci. Comput. Ser. **17**(4), 471–523 (1985)

---

[3] The service costs (or payment) can straightforwardly be extracted from text as they are simply numeric.

9. Comerio, M., Truong, H.-L., Paoli, F., Dustdar, S.: Evaluating contract compatibility for service composition in the SeCO$_2$ framework. In: Baresi, L., Chi, C.-H., Suzuki, J. (eds.) ICSOC/ServiceWave 2009. LNCS, vol. 5900, pp. 221–236. Springer, Heidelberg (2009). doi:10.1007/978-3-642-10383-4_15

10. Davey, B.A., Priestley, H.A.: Introduction to Lattices and Order. Cambridge University Press, Cambridge (2002)

11. Ferrari, G., Lluch-Lafuente, A.: A logic for graphs with QoS. Electron. Notes Theor. Comput. Sci. **142**, 143–160 (2006)

12. Pohle, G., Korsten, P.: Component business models - making specialization real. White paper. IBM®Institute for Business Value (2005)

13. Gabbay, D.M., Woods, J.: Logic and the Modalities in the Twentieth Century (Handbook of the History of Logic), vol. 7. North Holland, Amsterdam (2006)

14. Ghose, A., Lê, L.S., Hoesch-Klohe, K., Morrison, E.: The business service representation language: a preliminary report. In: Proceedings of the 1st International Workshop on Service Modelling and Representation Techniques - Associated With ServiceWave, Ghent, Belgium, December 2010

15. Governatori, G., Milosevic, Z.: A formal analysis of a business contract language. Int. J. Coop. Inf. Syst. **15**(4), 659–685 (2006)

16. Guizzardi, G.: On ontology, ontologies, conceptualizations, modeling languages, and (meta) models. In: Frontiers in Artificial Intelligence and Applications, Databases and Information Systems IV

17. Hansmann, U., Merk, L., Nicklous, M.S., Stober, T.: Pervasive Computing: The Mobile World, 2nd edn. Springer, Heidelberg (2011)

18. Hardouin, L., Cottenceau, B., Lhommeau, M., Le Corronc, E.: Interval systems over idempotent semiring. Linear Algebra Appl. **431**(5–7), 855–862 (2009)

19. Hinge, K., Ghose, A., Koliadis, G.: Process SEER: a tool for semantic effect annotation of business process models. In: Proceedings of the 13th IEEE International Conference on Enterprise Distributed Object Computing, pp. 49–58. IEEE Computer Society, Auckland, September 2009

20. Hirsch, D., Tuosto, E.: SHReQ: coordinating application level QoS. In: Proceedings of 3rd IEEE International Conference on Software Engineering and Formal Methods, Koblenz, Germany, pp. 425–434, September 2005

21. Holmes, T., Tran, H., Zdun, U., Dustdar, S.: Modeling human aspects of business processes – a view-based, model-driven approach. In: Schieferdecker, I., Hartman, A. (eds.) ECMDA-FA 2008. LNCS, vol. 5095, pp. 246–261. Springer, Heidelberg (2008). doi:10.1007/978-3-540-69100-6_17

22. IFM and IBM: succeeding through service innovation: a service perspective for education, research, business and government. White paper, University of Cambridge Institute for Manufacturing, Cambridge, UK (2008)

23. ISO/IEC: ITU-T X.902—ISO/IEC 10746-2 Information Technology - Open Distributed Processing - Reference Model - Foundations. International standard, SC 7 and ITU (2010)

24. Jackson, D.: Alloy: a lightweight object modelling notation. ACM Trans. Softw. Eng. Methodol. **11**(2), 256–290 (2002)

25. Kloppmann, M., Koenig, D., Leymann, F., Pfau, G., Rickayzen, A., Riegen, C., Schmidt, P., Trickovic, I.: WS-BPEL extension for people - BPEL4People. White paper, IBM and SAP, June 2005

26. Kohlborn, T., Luebeck, C., Korthaus, A., Fielt, E., Rosemann, M., Riedl, C., Krcmar, H.: Conceptualizing a bottom-up approach to service bundling. In: Pernici, B. (ed.) CAiSE 2010. LNCS, vol. 6051, pp. 129–134. Springer, Heidelberg (2010). doi:10.1007/978-3-642-13094-6_11

27. Kühne, T.: Matters of (meta-) modeling. Soft. Syst. Model. **5**, 369–385 (2006)
28. Lê, L.S.: Services for business processes in EA - are they in relation? In: Proceedings of the 22nd Australasian Conference on Information Systems. Association for Information Systems, Sydney, Australia, December 2011
29. Lê, L.-S.: Contractual proximity of business services. In: Dang, T.K., Wagner, R., Küng, J., Thoai, N., Takizawa, M., Neuhold, E. (eds.) FDSE 2015. LNCS, vol. 9446, pp. 183–197. Springer, Heidelberg (2015). doi:10.1007/978-3-319-26135-5_14
30. Lê, L.S., Dam, H., Ghose, A.: On business services representation -the $3 \times 3 \times 3$ approach. In: Proceedings of the 21st Australasian Conference on Information Systems. Association for Information Systems, Brisbane, Australia, December 2010
31. Lê, L.S., Ghose, A., Morrison, E.: Definition of a description language for business service decomposition. In: Proceedings of the 1st International Conference on Exploring Services Sciences (IESS 2010), Geneva, Switzerland, pp. 96–110, February 2010
32. Lê, L.S., Truong, H.L., Ghose, A., Dustdar, S.: On elasticity and constrainedness of business services provisioning. In: Proceedings of the 9th International Conference on Services Computing (SCC), pp. 384–391. IEEE Computer Society, June 2012
33. Lê, L.-S., Ghose, A.: Contracts + Goals = Roles? In: Atzeni, P., Cheung, D., Ram, S. (eds.) ER 2012. LNCS, vol. 7532, pp. 252–266. Springer, Heidelberg (2012). doi:10.1007/978-3-642-34002-4_20
34. Letier, E., van Lamsweerde, A.: Deriving operational software specifications from system goals. ACM SIGSOFT Softw. Eng. Notes **27**(6), 119–128 (2002)
35. Lieberman, H.: Using prototypical objects to implement shared behavior in object-oriented systems. In: Proceedings of the 1st Conference on Object-Oriented Programming Systems, Languages and Applications, pp. 214–223. ACM (1986)
36. Linington, P., Milosevic, Z., Cole, J., Gibson, S., Kulkarni, S., Neal, S.: A unified behavioural model and a contract language for extended enterprise. Data Knowl. Eng. **51**(1), 5–29 (2004)
37. Liskov, B.H., Wing, J.M.: A behavioral notion of subtyping. ACM Trans. Program. Lang. Syst. **16**(6), 1811–1841 (1994)
38. Liu, Y., Ngu, A.H., Zeng, L.Z.: QoS computation and policing in dynamic web service selection. In: Proceedings of the 13th International World Wide Web Conference on Alternate Track Papers and Posters - WWW Alt. 2004. Association for Computing Machinery (ACM) (2004)
39. Meyer, B.: Design By Contract. Prentice Hall, Englewood Cliffs (2005)
40. Paulson, L.D.: Services science: a new field for today's economy. Computer **39**(8), 18–21 (2006)
41. Raut, M., Singh, A.: Prime implicates of first order formulas. Int. J. Comput. Sci. Appl. **1**(1), 1–11 (2004)
42. Rosch, E.: Cognitive representations of semantic categories. J. Exp. Psychol. **104**(3), 192–233 (1975)
43. Saat, J., Franke, U., Lagerstrom, R., Ekstedt, M.: Enterprise architecture meta models for IT/business alignment situations. In: Proceedings of the 14th IEEE International Conference on Enterprise Distributed Object Computing, Vitória, Brazil, pp. 14–23, October 2010
44. Singh, M.P., Huhns, M.N.: Service-Oriented Computing: Semantics, Processes Agents. Wiley, Chichester (2005)
45. Smullyan, R.M.: First-Order Logic. Dover Publications, New York (1995)
46. Taivalsaari, A.: Classes vs. prototypes - some philosophical and historical observations. J. Object-Oriented Prog. **10**(7), 44–50 (1997)

47. Homann, U.: A business-oriented foundation for service orientation. In: White paper. Microsoft MSDN Library (2006)
48. Vu, L.-H., Hauswirth, M., Aberer, K.: QoS-based service selection and ranking with trust and reputation management. In: Meersman, R., Tari, Z. (eds.) OTM 2005. LNCS, vol. 3760, pp. 466–483. Springer, Heidelberg (2005). doi:10.1007/11575771_30

Energy-Saving Virtual Machine Scheduling in Cloud Computing with Fixed Interval Constraints

Nguyen Quang-Hung[✉], Nguyen Thanh Son, and Nam Thoai

Faculty of Computer Science and Engineering,
Ho Chi Minh City University of Technology,
VNU-HCM 268 Ly Thuong Kiet Street, Ho Chi Minh City, Vietnam
{hungnq2,sonsys,nam}@cse.hcmut.edu.vn

Abstract. Energy efficiency has become an important measurement of scheduling algorithms for Infrastructure-as-a-Service (IaaS) clouds. This paper investigates the energy-efficient virtual machine scheduling problems in IaaS clouds where users request multiple resources in fixed intervals and non-preemption for processing their virtual machines (VMs) and physical machines have bounded capacity resources. Many previous works are based on migration techniques to move on-line VMs from low utilization hosts and turn these hosts off to reduce energy consumption. However, the techniques for migration of VMs could not use in our case. The scheduling problem is NP-hard. Instead of minimizing the number used physical machines, we propose a scheduling algorithm EMinTRE-LDTF to minimize the sum of total busy time of all physical machines that is equivalent to minimize total energy consumption. In this paper, we present the proved approximation in general and special cases of the scheduling problem. Using Feitelson's and Lublin99's parallel workload models in the Parallel Workloads Archive, our simulation results show that algorithm EMinTRE-LDTF could reduce the total energy consumption compared with state-of-the-art algorithms including Tian's Modified First-Fit Decreasing Earliest, Beloglazov's Power-Aware Best-Fit Decreasing and Vector Bin-Packing Norm-based Greedy. Moreover, the EMinTRE-LDTF has less total energy consumption compared with our previous heuristic (e.g. MinDFT) in the simulations.

Keywords: Energy efficiency · Power-aware · Virtual machine · VM placement · VM allocation · IaaS · Scheduling · Cloud computing

1 Introduction

An Infrastructure-as-a-Service (IaaS) cloud system provides users with computing resources in terms of virtual machines (VMs) to run their applications [2,3,12,16,24]. These IaaS cloud systems are often built from virtualized data centers [2,3,24]. Power consumption in a large-scale data center requires multiple megawatts [8,16]. Le et al. [16] estimate the energy cost of a single data center

© Springer-Verlag GmbH Germany 2017
A. Hameurlain et al. (Eds.): TLDKS XXXI, LNCS 10140, pp. 124–145, 2017.
DOI: 10.1007/978-3-662-54173-9_6

is more than \$15M per year. As these data centers have more physical servers, they will consume more energy. Therefore, advanced scheduling techniques for reducing energy consumption of these cloud systems are highly concerned for any cloud providers to reduce energy cost. Increasing energy cost and the need to environmental sustainability address energy efficiency is a hot research topic in cloud systems. Energy-aware scheduling of VMs in IaaS cloud is still challenging [12, 16, 23, 25, 27].

Many previous works [3, 4, 20] proved that the virtual machine allocation is NP-hard and proposed to address the problem of energy-efficient scheduling of VMs in cloud data centers. They [3, 4, 20] present techniques for consolidating virtual machines in cloud data centers by using bin-packing heuristics (such as First-Fit Decreasing [20], and/or Best-Fit Decreasing [4]). They attempt to minimize the number of running physical machines and to turn off as many idle physical machines as possible. Consider a d-dimensional resource allocation where each user requests a set of virtual machines (VMs). Each VM requires multiple resources (such as CPU, memory, and IO) and a fixed quantity of each resource at a certain time interval. Under this scenario, using a minimum number of physical machines may not be a good solution. In a homogeneous environment where all physical servers are identical, the power consumption of each physical server is linear to its CPU utilization, i.e., a schedule with longer working time will consume more energy than another schedule with shorter working time.

Table 1. Example showing that using a minimum number of physical servers is not optimal. (*: demand resources are normalized to the maximum capacity resources of physical machines).

VM ID	CPU*	RAM*	Network*	Starttime	Dur. (hour)
VM1	0.5	0.1	0.2	0	10
VM2	0.5	0.5	0.2	0	2
VM3	0.2	0.4	0.2	0	2
VM4	0.2	0.4	0.2	0	2
VM5	0.1	0.1	0.1	0	2
VM6	0.5	0.5	0.2	1	9

Our work studies increasing time and resource efficiency-based approach to allocate VMs onto physical machines in other that it minimizes total energy consumption of all physical machines. Each VM requests resource allocation in a fixed starting time and non-preemption for the duration time. We present here an example to demonstrate our ideas to minimize total energy consumption of all physical machines in the VM placement with fixed starting time and duration time. For example, given six virtual machines (VMs) with their resource demands described in Table 1. Note that the maximum capacity of each resource is 1. In the example, a bin-packing-based algorithm could result in a schedule S_1 in which two physical servers are used: one for allocating VM1, VM3, VM4, and VM5; and another one for allocating VM2 and VM6. The resulted total completion time is $(10 + 10) = 20$ h. However, in another schedule S_2 in which where VMs are placed on three physical servers, VM1 and VM6 on the first physical server, VM3, VM4 and VM5 on the second physical server, and VM2 on the third physical server, then the total completion time of the five VMs is only $(10 + 2 + 2) = 14$ h.

This paper presents a proposed heuristic, denoted as EMinTRE-LDTF, to allocate VMs that request multiple resources in the fixed interval time and non-preemption into physical machines to minimize total energy consumption of physical machines while meeting all resource requirements. Using numerical simulations, we compare EMinTRE-LDTF with the state-of-the-art algorithms include Power-Aware Best-Fit Decreasing (PABFD) [4], vector bin-packing norm-based greedy (VBP-Norm-L2) [20], and Modified First-Fit-Decreasing-Earliest (Tian-MFFDE) [26]. Using two parallel workload models [9,17] in the Feitelson's Parallel Workloads Archive [10], our simulation results show that EMinTRE-LDTF could reduce the total energy consumption compared with PABFD [4], VBP-Norm-L2 [20], and Tian-MFFDE [26]. Moreover, EMinTRE-LDTF has less total energy consumption compared with MinDFT-LDTF [21] in the simulations.

The rest of this paper is structured as follows. Section 2 discusses related works. Section 3 describes the energy-aware VM allocation problem with multiple requested resources, fixed starting and duration time. We also formulate the objective of scheduling, and present our theorems. The proposed EMinTRE-LDTF algorithm present in Sect. 4. Section 5 discusses our performance evaluation using simulations. Section 6 concludes this paper and introduces future works.

2   Related Work

The interval scheduling problems have been studied for many years with objective to minimizing total busy time. In 2007, Kovalyov et al. [15] has presented work to describe characteristics of a fixed interval scheduling problem in which each job has fixed starting time, fixed processing time, and is only processed in the fixed duration time on a available machine. The scheduling problem can be applied in other domains. Angelelli et al. [1] considered interval scheduling with a resource constraint in parallel identical machines. The authors proved the decision problem is NP-complete if number of constraint resources in each parallel machine is a fixed number greater than two. Flammini et al. [11] studied using new approach of minimizing total busy time to optical networks application. Tian et al. [26] proposed a Modified First-Fit Decreasing Earliest algorithm, denoted as Tian-MFFDE, for placement of VMs energy efficiency. The Tian-MFFDE sorts list of VMs in queue order by longest their running times first) and places a VM (in the sorted list) to any first available physical machine that has enough VM's requested resources. Our VM placement problem differs from these interval scheduling problems [1,15,26], where each VM requires for multiple resource (e.g. computing power, physical memory, network bandwidth, etc.) instead of all jobs in the interval scheduling problems are equally on demanded computing resource (i.e. each physical machine can process the maximum of g jobs in concurrently).

Energy-aware resource management in cloud virtualized data centers is critical. Many previous research [3,4,7,14,25] proposed algorithms that consolidate VMs onto a small set of physical machines (PMs) in virtualized datacenters to minimize energy/power consumption of PMs. A group in Microsoft Research [20] has studied first-fit decreasing (FFD) based heuristics for vector bin-packing to minimize number of physical servers in the VM allocation problem. Some other works also proposed meta-heuristic algorithms to minimize the number of physical machines. Beloglazov et al. [3,4] have presented a modified best-fit decreasing heuristic in bin-packing problem, denoted as PABFD, to place a new VM to a host. PABFD sorts all VMs in a decreasing order of CPU utilization and tends to allocate a VM to an active physical server that would take the minimum increase of power consumption. Knauth et al. [14] proposed the OptSched scheduling algorithm to reduce cumulative machine up-time (CMU) by 60.1% and 16.7% in comparison to a round-robin and First-fit. The OptSched uses an minimum of active servers to process a given workload. In a heterogeneous physical machines, the OptSched maps a VM to a first available and the most powerful machine that has enough VM's requested resources. Otherwise, the VM is allocated to a new unused machine. In the VM allocation problem, however, minimizing the number of used physical machines is not equal to minimizing total of total energy consumption of all physical machines. Previous works do not consider multiple resources, fixed starting time and non-preemptive duration time of these VMs. Therefore, it is unsuitable for the power-aware VM allocation considered in this paper, i.g. these previous solutions can not result in a minimized total energy consumption for VM placement problem with certain interval time while still fulfilling the quality-of-service.

Chen et al. [7] observed there exists VM resource utilization patterns. The authors presented an VM allocation algorithm to consolidate complementary VMs with spatial and temporal-awareness in physical machines. They introduce resource efficiency and use norm-based greedy algorithm, which is similar to in [20], to measure distance of each used resource's utilization and maximum capacity of the resource in a host. Their VM allocation algorithm selects a host that minimizes the value of this distance metric to allocate a new VM. Our proposed EMinTRE-LDTF uses a different metric that unifies both increasing time and resource efficiency. In our proposed metric, the increasing time is the difference between two completion time of a host after and before allocating a VM.

Our proposed EMinTRE-LDTF algorithm that differs from these previous works. Our EMinTRE-LDTF algorithm use the VM's fixed starting time and duration to minimize the total busy time on physical machines, and consequently minimize the total energy consumption in all physical servers. To the best of our knowledge, no existing works that surveyed in [5,13,18,19] have thoroughly considered these aspects in addressing the problem of VM placement.

3   Problem Description

3.1   Notations

We use the following notations in this paper:

vm_i: The ith virtual machine to be scheduled.

M_j: The jth physical server.

S: A feasible schedule.

P^{idle}: The idle power consumption of a physical machine.

P^{max}: The maximum power consumption of a physical machine.

$P_j(t)$: The power consumption of M_j at a time point t.

ts_i: The fixed starting time of vm_i.

d_i: The duration time of vm_i.

T: The maximum schedule length, which is the time that the last virtual machine will be finished.

$\mathscr{J}_j$: The set of virtual machines that are allocated to M_j in the whole schedule.

T_j: The total busy time (working time) of M_j.

e_i: The energy consumption for running vm_i in the physical machine that vm_i is allocated.

g: The maximum number of virtual machines that can be assigned to any physical machine.

3.2   Problem Formulation

Consider the following scheduling problem. We are given a set of n virtual machines $\mathscr{V} = \{vm_1, \ldots, vm_n\}$ to be scheduled on a set of m identical physical servers $\mathscr{M} = \{M_1, \ldots, M_m\}$, each server can host a maximum number of g virtual machines. Each VM needs d-dimensional demand resources in a fixed interval with non-migration. Each vm_i is started at a fixed starting time (ts_i) and is non-preemptive during its duration time (d_i). Types of resource considered in the problem include computing power (i.e., the total Million Instruction Per Seconds (MIPS) of all cores in a physical machine), physical memory (i.e., the total MBytes of RAM in a physical machine), network bandwidth (i.e., the total Kb/s of network bandwidth in a physical machine), and storage (i.e., the total free GBytes of file system in a physical machine), etc.

The objective scheduling is to find out a feasible schedule S to minimize the total energy consumption of m physical servers. The objective scheduling is presented as:

$$\textbf{Minimize } (P^{idle} \times \sum_{j=1}^{m} T_j + \sum_{i=1}^{n} e_i) \tag{1}$$

where T_j is the total busy time of M_j. The $P^{idle} \times T_j$ is the minimum energy consumption of M_j, denoted as E_j^{min}, to keep it is on and active for during its total busy time (T_j), i.e., $E_j^{min} = P^{idle} \times T_j$. The $P^{idle} \times \sum_{j=1}^{m} T_j$ is sum of the minimum energy consumption of m used physical servers. The T_j is defined as the length of union of interval times of all VMs that are allocated to a physical machine M_j. Let $\mathscr{J}_j$ be set of virtual machines that are allocated to M_j in the whole schedule. T_j is defined as following:

$$T_j = len(\bigcup_{vm_i \in \mathscr{J}_j} [ts_i, ts_i + d_i]) \tag{2}$$

The scheduling problem has the following hard constraints, which are firstly described in our previous work [21], as following:

- Constraint 1: Each VM is only processed by a physical server at any time with non-migration and non-preemption.
- Constraint 2: Each VM does not request any resource larger than the maximum total capacity resource of any physical server.
- Constraint 3: The sum of total demand resources of these allocated VMs is less than or equal to the total capacity of the resources of M_j. Each VM is represented as a d-dimensional vector of demand resources, i.e. $vm_i = (x_{i,1}, x_{i,2}, ..., x_{i,d})$. Similarly, each physical machine is denoted as a d-dimensional vector of capacity resources, i.e. $M_j = (y_{j,1}, y_{j,2}, ..., y_{j,d})$. Thus we have $\forall r \in \{1, ..., d\}, i \in \{1, 2, ..., n\}, j \in \{1, 2, ..., m\}$:

$$\sum_{vm_i \in \mathscr{J}_j} x_{i,r} \leq y_{j,r} \tag{3}$$

where:

- $x_{i,r}$ is resource of type r (e.g. CPU core, computing power, memory, etc.) requested by the vm_i (i = 1, 2,..., n).
- $y_{j,r}$ is capacity resource of type r (e.g. CPU core, computing power, memory, etc.) of the physical machine M_j ($j = 1, 2, ..., m$).

With at least one type of resource (i.e., $d \geq 1$), the scheduling problem is NP-hard [20].

3.3   Power Consumption Model

In this paper, we use the following energy consumption model proposed in [8] for a physical machine. The power consumption of M_j, denoted as $P_j(.)$, is formulated as follow $\forall j \in \{1, 2, ..., m\}$:

$$P_j(t) = P^{idle} + (P^{max} - P^{idle})U_j(t) \tag{4}$$

in which $U_j(t)$ is the CPU utilization of M_j at time t, P^{idle} and P^{max} are the idle power and the maximum power consumed at 0% and 100% CPU utilization respectively of a physical machine (all physical machines are homogeneous).

We assume that all cores in CPU are homogeneous, i.e. $\forall c = 1, 2, ..., PE_j$: $MIPS_{j,c} = MIPS_{j,1}$. The $U_j(t)$ is formulated as follow:

$$U_j(t) = (\frac{1}{PE_j \times MIPS_{j,1}}) \sum_{c=1}^{PE_j} \sum_{vm_i \in \mathscr{J}_j} mips_{i,c} \tag{5}$$

The energy consumption of M_j in the time period $[t_1, t_2]$ is formulated as follow:

$$E_j = \int_{t_1}^{t_2} P_j(U_j(t))dt \tag{6}$$

where:

$U_j(t)$: The CPU utilization of M_j at time t and $0 \leq U_j(t) \leq 1$.

PE_j: The number of processing elements (i.e. cores) of M_j.

$MIPS_{j,c}$: The maximum total computing power (in MIPS) of cth processing element on M_j.

$mips_{i,c}$: The allocated MIPS of the cth processing element to vm_i by M_j.

3.4   Preliminaries

Definition 1 (Length of intervals). *Given a time interval* $I = [s, f]$, *the length of* I *is* $len(I) = f - s$. *Extensively, to a set* $\mathscr{I}$ *of intervals, length of* $\mathscr{I}$ *is* $len(\mathscr{I}) = \sum_{I \in \mathscr{I}} len(I)$.

Definition 2 (Span of intervals). *For a set* $\mathscr{I}$ *of intervals, we define the span of* $\mathscr{I}$ *as* $span(\mathscr{I}) = len(\bigcup \mathscr{I})$.

Definition 3 (Optimal schedule). *An optimal schedule is the schedule that minimizes the total busy time of physical machines. For any instance* $\mathscr{J}$ *and parameter* $g \geqslant 1$, $OPT(\mathscr{J}, g)$ *denotes the cost of an optimal schedule.*

In this paper, we denote $\mathscr{J}$ is set of time intervals that derived from given set of all requested VMs. In general, we use instance $\mathscr{J}$ is alternative meaning to a given set of all requested VMs in context of this paper.

Observations: Cost, capacity, span bounds. For any instance $\mathscr{J}$, which is set of time intervals derived from given set of all requested VMs, and capacity parameter $g \geqslant 1$, which is the maximum number of VMs that can be allocated on any physical machine, the following bounds are held:

- The optimal cost bound: $OPT(\mathscr{J}, g) \leq len(\mathscr{J})$.
- The capacity bound: $OPT(\mathscr{J}, g) \geqslant \dfrac{len(\mathscr{J})}{g}$.
- The span bound: $OPT(\mathscr{J}, g) \geqslant span(\mathscr{J})$.

For any feasible schedule s on a given set of virtual machines, the total busy time of all physical machines that are used in the schedule s is bounded by the maximum total length of all time intervals in a given instance $\mathscr{J}$. Therefore, the optimal cost bound holds because $OPT(\mathscr{J}, g) = len(\mathscr{J})$ iff all intervals are non-overlapping, i.e., $\forall I_1, I_2 \in \mathscr{J}$ then $I_1 \cap I_2 = \emptyset$.

Intuitively, the capacity bound holds because $OPT(\mathscr{J}, g) = \dfrac{len(\mathscr{J})}{g}$ iff, for each physical server, exactly g VMs are neatly scheduled in that physical server. The span bound holds because at any time $t \in \bigcup \mathscr{J}$ at least one machine is working.

3.5   Theorems

Theorem 1. *Given a cloud system with a set of identical physical machines, assume that the power consumption of a physical machine is $P(u) = P^{idle} + (P^{max} - P^{idle})u$, where P^{idle} is the idle power consumption, P^{max} is the maximum power consumption, and u is the CPU utilization in percentage ($0 \le u \le 1$). We denote e_{ij} is energy consumption of the virtual machine i^{th} that is scheduled or mapped on the physical machine j^{th}. If the utilization u of the mapped virtual machine is a constant, then the energy consumption of each virtual machine, e_{ij}, is independent of any mapping (i.e. any schedule). We have $\forall i \in \{1, \ldots, n\}, j \in \{1, \ldots, m\} : e_{ij} = e_i$.*

Proof. Recall that the energy consumption is formulated in Eq. (6), and power consumption, $P(u)$, is a linear function of CPU utilization, u. Therefore $\forall i \in \{1, \ldots, n\}, j \in \{1, \ldots, m\}$, we see that e_{ij} is the integral of the $P(u)$ over any time interval $[t_1, t_2]$, and is the same value, denoted as e_i.

From Theorem 1, we can imply the following theorem.

Theorem 2. *Minimizing total energy consumption in (1) is equivalent to minimizing the sum of total busy time of all physical machines ($\sum_{j=1}^{m} T_j$).*

$$\textbf{\textit{Minimize }} (P^{idle} \times \sum_{j=1}^{m} T_j + \sum_{i=1}^{n} e_i) \sim \textbf{\textit{Minimize }} (\sum_{j=1}^{m} T_j) \qquad (7)$$

Proof. According to the objective function described in (1), P^{idle} is constant while e_i is independent of any mapping (i.e. any schedule).

Based on the above observations, we propose our energy-aware algorithms denoted as EMinTRE-LDTF which is presented in the next section.

Definition 4. *For any schedule we denote by $\mathscr{J}_j$ the set of virtual machines allocated to the physical machine M_j by the schedule. Let T_j denote the total busy time of M_j is the span of $\mathscr{J}_j$, i.e., $T_j = span(\mathscr{J}_j)$.*

Definition 5. *For any instance $\mathscr{J}$, the total busy time of the entire schedule of $\mathscr{J}$ computed by the algorithm H, which denoted as $cost^H(\mathscr{J})$, is defined as:*

$$cost^H(\mathscr{J}) = \int_0^{span(\mathscr{J})} N^H(t)dt \qquad (8)$$

in which $N^H(t)$ is the number of physical machines used at the time t by the algorithm H.

Definition 6. *For any instance $\mathcal{J}$ and parameter $g \geqslant 1$, $E^{OPT}(\mathcal{J},g)$, which is denoted as the minimized total energy consumption of all physical machines in an optimal schedule for the $\mathcal{J}$, is formulated as:* $E^{OPT}(\mathcal{J},g) = P^{idle} \cdot OPT(\mathcal{J},g) + \sum_{i=1}^{n} e_i$.

Theorem 3. *For any instance $\mathcal{J}$, the lower and upper of the total energy consumption in an optimal schedule are bounded by:* $P^{idle} \cdot \frac{len(\mathcal{J})}{g} \leq E^{OPT}(\mathcal{J},g) \leq P^{max} \cdot len(\mathcal{J})$.

Proof. For any instance $\mathcal{J}$, let $OPT(\mathcal{J},g)$ be the total busy time of the optimal schedule for the $\mathcal{J}$, and let E^* be the total energy consumption for the optimal schedule for the $\mathcal{J}$.

The total energy consumption of an optimal schedule needs to account for all physical machines running during $OPT(\mathcal{J},g)$. We have: $E^* = P^{idle} \cdot OPT(\mathcal{J},g) + \sum_{i=1}^{n} e_i$.

From Definition 6, we have $E^{OPT}(\mathcal{J},g) = E^*$.

Apply the capacity bound in Theorem 3.4, we have $OPT(\mathcal{J},g) \geq \frac{len(\mathcal{J})}{g}$.

Thus, $E^* \geq P^{idle} \cdot \frac{len(\mathcal{J})}{g} + \sum_{i=1}^{n} e_i$.

Recall that the energy consumption of each virtual machine is non-negative, thus $e_i > 0$. Therefore, $E^* \geq P^{idle} \cdot \frac{len(\mathcal{J})}{g}$. Thus

$$E^{OPT}(\mathcal{J},g) \geq P^{idle} \cdot \frac{len(\mathcal{J})}{g} \tag{9}$$

We prove the upper bound of the minimized total energy consumption as following. Apply the optimal cost bound in the Observations, we have $OPT(\mathcal{J},g) \leq len(\mathcal{J})$.

Thus

$$E^* \leq P^{idle} \cdot len(\mathcal{J}) + \sum_{i=1}^{n} e_i. \tag{10}$$

Apply the linear power consumption as in the Eqs. (4) and (6), the energy consumption of each i-th virtual machine in period time of $[ts_i, ts_i + d_i]$ that denotes as e_i is:

$$e_i = \int_{ts_i}^{ts_i+d_i} P_j(U_{vm_i})dt = (P_j^{max} - P_j^{idle}) \cdot U_{vm_i} \cdot d_i$$

where U_{vm_i} is the percentage of CPU usage of the i-th virtual machine on a j-th physical machine.

Because any virtual machine always requests CPU usage lesser than or equal to the maximum total capacity CPU of every physical machine, i.e., $U_{vm_i} \leq 1$.

$$\Rightarrow e_i \leq (P_j^{max} - P_j^{idle}) \cdot d_i$$

Note that in this proof, all physical machines are identical with same power consumption model thus P^{max} and P^{idle} are the maximum power consumption and the idle power consumption of each physical machine. Thus:

$$e_i \leq (P^{max} - P^{idle}) \cdot d_i$$

Let I_i is interval of each i-th virtual machine, $I_i = [ts_i, ts_i + d_i]$. By the definition the length of interval is $len(I_i) = d_i$ that is duration time of each i-th virtual machine. Thus:

$$e_i \leq (P^{max} - P^{idle}) \cdot len(I_i)$$

The total energy consumption of n virtual machines is formulated as:

$$\sum_{i=1}^{n} e_i \leq \sum_{i=1}^{n} [(P^{max} - P^{idle}) \cdot len(I_i)] \Leftrightarrow \sum_{i=1}^{n} e_i \leq (P^{max} - P^{idle}) \cdot \sum_{i=1}^{n} len(I_i)$$

$$\Leftrightarrow \sum_{i=1}^{n} e_i \leq (P^{max} - P^{idle}) \cdot len(\mathscr{J}). \tag{11}$$

From Equation (10), we have:

$$E^* \leq P^{idle} \cdot len(\mathscr{J}) + \sum_{i=1}^{n} e_i E^* \leq P^{idle} \cdot len(\mathscr{J}) + (P^{max} - P^{idle}) \cdot len(\mathscr{J})$$

$$E^* \leq (P^{idle} + (P^{max} - P^{idle})) \cdot len(\mathscr{J}) \tag{12}$$

From the Equation (12):

$$E^* \leq P^{max} \cdot len(\mathscr{J}) \tag{13}$$

$$\Leftrightarrow E^{OPT}(\mathscr{J}, g) \leq P^{max} \cdot len(\mathscr{J}) \tag{14}$$

From both of two Eqs. (9) and (14), we have:

$$P^{idle} \cdot \frac{len(\mathscr{J})}{g} \leq E^{OPT}(\mathscr{J}, g) \leq P^{max} \cdot len(\mathscr{J}) \tag{15}$$

We prove the theorem.

4 Scheduling Algorithm

4.1 EMinTRE-LDTF Scheduling Algorithm

In this section, we present our energy-aware scheduling algorithm, namely, EMinTRE-LDTF. EMinTRE-LDTF presents a metric to unify the increasing time and estimated resource efficiency when mapping a VM onto a physical machine. Then, EMinTRE-LDTF will choose a host that minimizes the metric. Our previous MinDFT-LDTF and MinDFT-LFT, which use core algorithm MinDFT in [21], only focused on minimizing the increasing time when mapping a VM onto a physical machine. MinDFT-LDTF sorts the list of VM oder by longest duration time first, and MinDFT-LFT sorts the list of VM oder by latest finishing time first. Algorithm EMinTRE-LDTF additionally considers resource efficiency during an execution period of a physical machine in order to fully utilize resources in a physical machine. Algorithm EMinTRE-LDTF differs from EMinRET [22] in the equation of TRE and EMinTRE-LDTF does not have swapping step as in EMinRET.

Based on Eq. 5, the utilization of a resource r (resource r can be CPU, physical memory, network bandwidth, storage, etc.) of the M_j, denoted as $U_{j,r}$, is formulated as:

$$U_{j,r} = \sum_{s \in n_j} \frac{V_{s,r}}{H_{j,r}}. \tag{16}$$

where n_j is the list of virtual machines that are assigned to the M_j, $V_{s,r}$ is the amount of requested resource r of the virtual machine s (note that in our study the value of $V_{s,r}$ is fixed for each user request), and $H_{j,r}$ is the maximum capacity of the resource r in M_j.

Inspired by the work from Microsoft research team [7,20], the resource efficiency of a physical machine j^{th}, denoted by RE_j, is Norm-based distant [20] of two vectors: normalized resource utilization vector and unit vector $\mathbf{1}$. The resource efficiency is formulated as:

$$RE_j = \sum_{r \in \mathcal{R}} ((1 - U_{j,r}) \times w_r)^2 \tag{17}$$

where R is the set of resource types in a host ($\mathcal{R} = \{$cpu, ram, netbw, io, storage$\}$) and w_r is weight of resource r in a physical machine.

In this paper, we propose a unified metric for increasing time and resource efficiency of the host j-th that is calculated as:

$$TRE_j = \begin{cases} \sqrt{RE_j}, & \text{if } t^{diff} = 0. \\ (\frac{t^{diff}}{3600} \times w_{r=time})\sqrt{RE_j}, & \text{if } t^{diff} \neq 0. \end{cases} \tag{18}$$

EMinTRE-LDTF chooses a physical host that has a minimum value of the TRE metric to allocate for a VM. We present the pseudo-code of EMinTRE-LDTF in Algorithm 1. EMinTRE-LDTF has two (2) steps: firstly, EMinTRE-LDTF sorts the list of VMs by longest duration time first, and secondly,

Algorithm 1. EMinTRE-LDTF: Energy-Aware Minimizing Resource Efficiency - Time

```
1: function EMINTRE-LDTF
2:     Input: vmList - a list of virtual machines to be scheduled, hostList - a list of physical
       servers
3:     Output: a feasible schedule or null
4:     vmList = sortVmListByOrderLongestDurationTimeFirst( vmList )          ▷ 1
5:     m = hostList.size(); n = vmList.size();
6:     T[j] = 0, ∀j ∈ [1, m]
7:     for i = 1 to n do                                                     ▷ on the VMs list
8:         vm = vmList.get(i)
9:         allocatedHost = null
10:        T1 = sumTotalHostCompletionTime( T )
11:        minTRE = +∞
12:        for j = 1 to m do                                                 ▷ on the hosts list
13:            host = hostList.get( j )
14:            hostVMList = sortVmListByOrder( host.getVms(), order=[starttime, finishtime])
15:            if host.checkAvailableResource( vm ) then
16:
17:                preTime = T[ host.id ]
18:                T[ host.id ] = host.estimateHostTotalCompletionTime( vm )
19:                T2 = sumTotalHostCompletionTime( T )
20:                diffTime = Math.max( T2 - T1, 0)
21:                TRE = EstimateMetricTimeResEff( diffTime, host )
22:                if (minTRE > TRE ) then
23:                    minTRE = TRE
24:                    allocatedHost = host
25:                end if
26:                T[ host.id ] = preTime       ▷ Next iterate over the hostList and choose the host
           that minimize the value of different time and resource efficiency
27:            end if
28:        end for
29:        if (allocatedHost != null) then
30:            allocate the vm to the host
31:            add the pair of vm (key) and host to the mapping
32:        end if
33:    end for
34:    return mapping
35: end function
36: sumTotalHostCompletionTime(T[]) = ∑_{j=1}^{m} T_j      ▷ T[1...m]: Array of total completion times
       of m physical servers
```

EMinTRE-LDTF schedule the first VM in the sorted list of VMs to a host that has the minimum of the TRE. The EMinTRE-LDTF solves the scheduling problem in time complexity of $\mathcal{O}(n \times m \times q)$ where n is the number of VMs to be scheduled, m is the number of physical machines, and q is the maximum number of allocated VMs in the physical machines $M_j, \forall j = 1, 2, ..., m$.

4.2 Approximation Algorithm for General Case

In this section, we claim that algorithm EMinTRE-LDTF for general instance yields its approximation ratio are g, where g is the maximal number of virtual

---

Algorithm 2. Estimating the metric for increasing time and resource efficiency

---

1: **function** ESTIMATEMETRICTIMERESEFF
2: ＿ *Input:* $(t^{diff}, host)$ - t^{diff} is a different time, *host* is a candidate physical machine
3:     *Output: TRE* - a value of metric time and resource efficiency
4:     Set $\mathscr{R}=\{$cpu, ram, netbw, io, storage, time$\}$
5:     $j = $ host.getId(); $n_j = $ host.getVMList();
6:     **for** $r \in \mathscr{R}$ **do**
7:         Calculate the resource utilization, $U_{j,r}$ as in the Equaltion (16).
8:     **end for**
9:     $weights[] \leftarrow$ Read resource weights from configuration file
10:     Calculate the different time and resource efficiency metric for host j denoted as TRE_j
    as in the Equaltion (18)
11:     **return** TRE
12: **end function**

---

machines can be assigned to each physical machine. EMinTRE-LDTF sorts the list of virtual machines in order of their longest duration time first.

Theorem 4. *For any instance $\mathscr{J}$, EMinTRE-LDTF is a g-approximation algorithm where g is the maximal number of virtual machines can be assigned to each physical machine, i.e. the total busy time of schedule for $\mathscr{J}$ outputted by EMinTRE-LDTF is the maximum g times the total busy time of optimal schedule. We denote $EMinTRE - LDTF(\mathscr{J})$ is cost of algorithm EMinTRE-LDTF for a given instance $\mathscr{J}$ that is defined in the Definition 5. Formally,*

$$EMinTRE - LDTF(\mathscr{J}) \leq g \cdot OPT(\mathscr{J}). \tag{19}$$

Proof. Let $N(t)$ denote the number of virtual machines that could be placed at time t. Let $N^H(t)$ denote the number of used physical machines at time t in a schedule that is resulted by algorithm EMinTRE-LDTF, and $N^{OPT}(t)$ denotes the number of machines used at time t in an optimal schedule.

For any time $t \geq 0$, each of using $N^H(t)$ physical machines has at least one allocated virtual machine, i.e., $N(t) \geq N^H(t)$. Clearly, at any time $t \geq 0$ and $g \geq 1$, $N^{OPT}(t) \geq \frac{N(t)}{g} = \frac{N^H(t)}{g}$.

The total busy time of the entire schedule of an instance $\mathscr{J}$ denoted as $EMinTRE - LDTF(\mathscr{J})$ is calculated by taking the integral of $N^H(t)$ with all values of $t \in [0, span(\mathscr{J})]$. Thus, we have:

$$N^{OPT}(t) \geq \frac{N^H(t)}{g} \Longleftrightarrow N^H(t) \leq g \cdot N^{OPT}(t)$$

$$\Rightarrow \int_0^{span(\mathscr{J})} N^H(t)dt \leq g \cdot \int_0^{span(\mathscr{J})} N^{OPT}(t)dt$$

$$\Longleftrightarrow EMinTRE - LDTF(\mathscr{J}) \leq g \cdot OPT(\mathscr{J})$$

Thus, by applying the same reasoning to other algorithms such as EMinTRE-LDTF, we have:

$$EMinTRE - LDTF(\mathscr{J}) \leq g \cdot OPT(\mathscr{J})$$

4.3   Approximations for Special Cases

Proper Interval Graphs. In this section we consider instances in which no virtual machine's time interval is properly contained in another. The intersection graphs for such instances are known as *proper interval graphs*. Algorithm EMinTRE-LDTF for proper interval graphs includes two steps. In the first step, the list of virtual machines is sorted by their earliest starting time first. In the second step, each virtual machine is placed at the currently filled physical machine so that TRE metric of the physical machine is minimized, unless the placement violates the hard constraint on capacity of the physical machine, in which case a new physical machine is opened.

Theorem 5. *Algorithm EMinTRE-LDTF yields a $(3 - \frac{2}{g})$ for proper interval graphs, where g is the maximum number of virtual machines that could be placed on a physical machine in satisfying all their resource requirement*

Proof. Let $N(t)$ denote the number of virtual machines that could be placed at time t. Let $N^H(t)$ denote the number of used physical machines at time t in a schedule that is resulted by algorithm EMinTRE-LDTF (H), and $N^{OPT}(t)$ denotes the number of machines used at time t in an optimal schedule (OPT).

Theorem 6. (Proposition). *For any t, $N(t) \geq (N^H(t) - 2)g + 2$.*

Proof. For a given $t > 0$, let $m = N^H(t)$ denote number of used physical machines at time t. There are $m - 2$ additional used machines denote as $M_2, ..., M_{m-1}$. The first machine M_1 has at least one virtual machine that is placed on the M_1, and other machines $M_2, ..., M_{m-1}$ are assigned fully g virtual machines on each, and the m-th machine M_m is assigned at least one virtual machine. Since the graph is proper, suppose that the first machine processes at time t one virtual machine u, any virtual machine v is placed to another machine starts after the u and ends after u, thus v is running at time t. With m used physical machines at time t, the number of running virtual machines at time t is at least $(m - 2)g + 2$. Therefore $N(t) \geq (m - 2)g + 2$.

For any t, $N^{OPT}(t) \geq \frac{N(t)}{g}$. Applying the proposition, we have at any t, $N(t) \geq (N^H(t) - 2)g + 2$

$$N^{OPT}(t) \geq \frac{(N^H(t) - 2)g + 2}{g}$$

Recall $g > 0$, thus

$$N^H(t) \leq N^{OPT}(t) + 2 - \frac{2}{g}$$

The total busy time of entire schedule of $\mathscr{J}$ is:

$$EMinTRE - LDTF(\mathscr{J}) = \int_0^{span(\mathscr{J})} N^H(t)dt$$

$$EMinTRE - LDTF(\mathscr{J}) \le \int_0^{span(\mathscr{J})} (N^{OPT}(t) + 2 - \frac{2}{g})dt$$

$$EMinTRE - LDTF(\mathscr{J}) \le OPT(\mathscr{J}) + (2 - \frac{2}{g}) \cdot span(\mathscr{J}) \qquad (20)$$

With related to the span bound in Definition 5, we have $OPT(\mathscr{J}) \ge span(\mathscr{J})$ thus inequality (20) equivalent to:

$$EMinTRE - LDTF(\mathscr{J}) \le OPT(\mathscr{J}) + (2 - \frac{2}{g}) \cdot OPT(\mathscr{J})$$

$$\frac{EMinTRE - LDTF(\mathscr{J})}{OPT(\mathscr{J})} \le 3 - \frac{2}{g} \qquad (21)$$

This gives the statement of the Theorem 5.

Theorem 7. *If all virtual machines are homogeneous and each physical machine processes only one virtual machine then the EMinTRE-LDTF algorithm yields an optimal solution for proper interval graphs.*

Proof. In case all virtual machines are homogeneous and each physical machine processes only one virtual machine, thus $k = 1$ and $g = 1$. We have $EMinTRE - LDTF(\mathscr{J}) \ge OPT(\mathscr{J})$ and apply Theorem 5, thus $EMinTRE - LDTF(\mathscr{J}) = OPT(\mathscr{J})$.

5   Performance Evaluation

5.1   Algorithms

In this section, we study the following VM allocation algorithms:

- PABFD, a power-aware and modified best-fit decreasing heuristic [3,4]. The PABFD sorts the list of VM_i (i=1, 2,..., n) by their total requested CPU utilization, and assigns new VM to any host that has a minimum increase in power consumption.
- VBP-Norm-L2, a vector packing heuristics that is presented as Norm-based Greedy with degree 2 [20]. Weights of these Norm-based Greedy heuristics use FFDAvgSum which are $exp(x)$, which is the value of the exponential function at the point x, where x is average of sum of demand resources (e.g. CPU, memory, storage, network bandwidth, etc.). VBP-Norm-L2 assigns new VM to any host that has minimum of these norm values.
- MinDFT-LDTF and MinDFT-LFT: core of both MinDFT-LDTF and MinDFT-LFT algorithms is MinDFT [21]. MinDFT-LDTF sorts the list of VM_i (i = 1, 2,..., n) by their starting time (ts_i) and respectively by their finished time ($ts_i + dur_i$), then MinDFT-LDTF allocates each VM (in a given sorted list of VMs) to a host that has a minimum increase in total completion times of hosts. MinDFT-LFT differs MinDFT-LDTF in sorting the list of VMs, MinDFT-LFT sorts the list of VMs by their respectively finished time in latest finishing time first.

– EMinTRE-LDTF, the algorithm is proposed in the Sect. 4. EMinTRE-LDTF sorts the list of VMs (input) by VM's longest duration time first and host's allocated VMs by its finishing time and place a VM to any physical machine that minimizes time-resource efficiency (TRE) metric.

5.2 Methodology

We evaluate these algorithms by simulation using the CloudSim [6] to create a simulated cloud data center system that has identical physical machines, heterogeneous VMs, and with thousands of CloudSim's cloudlets [6] (we assume that each HPC job's task is modeled as a cloudlet that is run on a single VM). The information of VMs (and also cloudlets) in these simulated workloads is extracted from two parallel job models are Feitelson's parallel workload model [9] and Lublin99's parallel workload model [17] in Feitelson's Parallel Workloads Archive (PWA) [10]. When converting from the generated log-trace files, each cloudlet's length is a product of the system's processing time and CPU rating (we set the CPU rating is equal to included VM's MIPS). We convert job's submission time, job's start time (if the start time is missing, then the start time is equal to sum of job's submission time and job's waiting time), job's request run-time, and job's number of processors in job data from the log-trace in the PWA to VM's submission time, starting time and duration time, and number of VMs (each VM is created in round-robin in the four types of VMs in Table 2 on the number of VMs). Eight (08) types of VMs as presented in the Table 2 are used in the [26] that are similar to categories in Amazon EC2's VM instances: high-CPU VM, high-memory VM, small VM, and micro VM, etc. All physical machines are identical and each physical machine is a typical physical machine (Hosts) with 16 cores CPU (3250 MIPS/core), 136.8 GBytes of available physical memory, 10 Gb/s of network bandwidth, 10 TBytes of available storage. Power model of each physical machine is 175 W at idle power and 250 W at maximum power consumption (the idle power is 70% of the maximum power consumption as in [3,4,8]). In the simulations, we use weights as following: (i) weight of increasing time of mapping a VM to a host: {0.001, 0.01, 1, 100, 3600}; (ii) all weights of computing resources (e.g. number of MIPS per CPU core, physical

Table 2. Eight (08) VM types in simulations.

VM type	MIPS	Cores	Memory (Unit: MBytes)	Network (Unit: Mbits/s)	Storage (Unit: GBytes)
Type 1	2500	8	6800	100	1000
Type 2	2500	2	1700	100	422.5
Type 3	3250	8	68400	100	1000
Type 4	3250	4	34200	100	845
Type 5	3250	2	17100	100	422.5
Type 6	2000	4	15000	100	1690
Type 7	2000	2	7500	100	845
Type 8	1000	1	1875	100	211.25

Table 3. Information of a typical physical machine (host) with 16 cores CPU (3250 MIPS/core), 136.8 GBytes of available physical memory, 10 Gb/s of network bandwidth, 10 TBytes of storage and idle, maximum power consumption is 175, 250 (W).

Type	MIPS	Cores	Memory (Unit: MBytes)	Network (Unit: Mbits/s)	Storage (Unit: GBytes)	P^{idle} (Unit: Watts)	P^{max} (Unit: Watts)
M1	3250	16	140084	10000	10000	175	250

Table 4. The normalized total energy consumption. Simulation results of scheduling algorithms solving scheduling problems with 12681 VMs and 5000 physical machines (hosts) using Feiltelson's parallel workload model [9].

Algorithms	#Hosts	#VMs	Energy (KWh)	Norm. energy
PABFD	5000	12681	$1,055.42$	1.598
VBP-Norm-L2	5000	12681	$1,054.69$	1.597
Tian-MFFDE	5000	12681	660.30	1.000
MinDFT-LDTF	5000	12681	603.90	0.915
MinDFT-LFT	5000	12681	503.43	0.762
EMinTRE-LDTF	5000	12681	496.55	0.752

Table 5. The normalized total energy consumption. Simulation results of scheduling algorithms solving scheduling problems with 29,177 VMs and 10,000 physical machines (hosts) using Feiltelson's parallel workload model [9].

Algorithms	#Hosts	#VMs	Energy (KWh)	Norm. energy
PABFD	10000	29177	2261.878	1.540
VBP-Norm-L2	10000	29177	2260.615	1.539
Tian-MFFDE	10000	29177	1468.409	1.000
MinDFT-LDTF	10000	29177	1373.764	0.936
MinDFT-LFT	10000	29177	1109.852	0.756
EMinTRE-LDTF	10000	29177	1092.365	0.744

memory (RAM), network bandwidth, and storage) are equally to 1. We simulate on combination of these weights. The total energy consumption of each EMinTRE-LDTF is the average of five times simulation with various weights of increasing time (e.g. 0.001, 0.01, 1, 100, or 3600) (Tables 4, 5, 6 and 7).

We choose Modified First-Fit Decreasing Earliest (denoted as Tian-MFFDE) [26] as the baseline because Tian-MFFDE is the best algorithm in the energy-aware scheduling algorithm to time interval scheduling. We also compare our proposed VM allocation algorithms with PABFD [4] because the PABFD is a famous power-aware best-fit decreasing in the energy-aware scheduling research community, and two vector bin-packing algorithms (VBP-Norm-L1/L2) to show the importance of with/without considering VM's starting time and finish time in reducing the total energy consumption of VM placement problem.

Table 6. The normalized total energy consumption. Simulation results of scheduling algorithms solving scheduling problems with 8847 VMs and 5000 physical machines (hosts) using Lublin99's parallel workload model [17].

Algorithms	#Hosts	#VMs	Energy (KWh)	Norm. energy
PABFD	5000	8847	460.664	1.601
VBP-Norm-L2	5000	8847	453.229	1.575
Tian-MFFDE	5000	8847	287.779	1.000
MinDFT-LDTF	5000	8847	263.860	0.917
MinDFT-LFT	5000	8847	232.286	0.807
EMinTRE-LDTF	5000	8847	220.675	0.767

Table 7. The normalized total energy consumption. Simulation results of scheduling algorithms solving scheduling problems with 19853 VMs and 5000 physical machines (hosts) using Lublin99's parallel workload model [17].

Algorithms	#Hosts	#VMs	Energy (KWh)	Norm. energy
PABFD	5000	19853	3107.78	1.424
VBP-Norm-L2	5000	19853	3106.56	1.423
Tian-MFFDE	5000	19853	2182.54	1.000
MinDFT-LDTF	5000	19853	1927.52	0.883
MinDFT-LFT	5000	19853	1746.12	0.800
EMinTRE-LDTF	5000	19853	1485.13	0.680

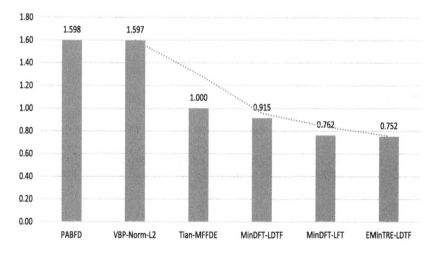

Fig. 1. The normalized total energy consumption compare to Tian-MFFDE. Result of simulations with Feitelson's Parallel Workload Archive Model [9] that includes 1,000 jobs have total of 12,681 VMs.

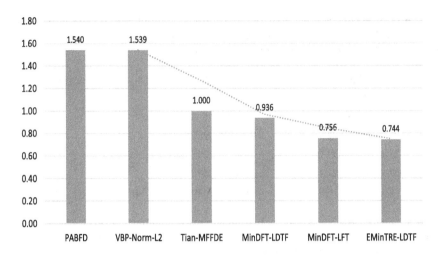

Fig. 2. The normalized total energy consumption compare to Tian-MFFDE. Result of simulations with Feitelson's parallel workload model [9] that includes 2,000 jobs have total of 29,177 VMs.

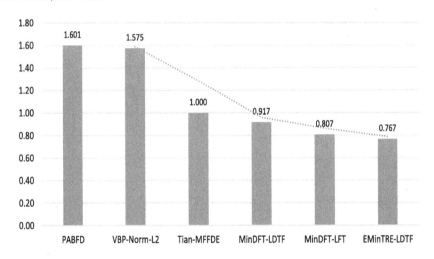

Fig. 3. The normalized total energy consumption compare to Tian-MFFDE. Result of simulations with Lublin99's parallel workload model [17] that includes 1,000 jobs have total of 8,847 VMs.

5.3   Results and Discussions

The Tables 4 and 5 show simulation results of scheduling algorithms solving scheduling problems with 12681 VMs - 5000 physical machines (hosts) and 29,177 VMs - 10,000 physical machines (hosts), in which VM's data is converted from the Feiltelson's parallel workload model [9] with 1000 jobs and 2000 jobs. The Tables 6 and 7 show simulation results of scheduling algorithms solving schedul-

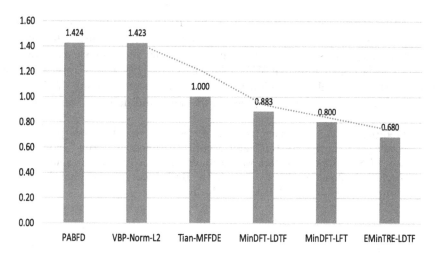

Fig. 4. The normalized total energy consumption compare to Tian-MFFDE. Result of simulations with Lublin99's parallel workload model [17] that includes 2,000 jobs have total of 19,853 VMs.

ing problems with 8847 VMs - 5000 physical machines (hosts) and 19853 VMs - 5000 physical machines (hosts), in which VM's data is converted from the Lublin99's parallel workload model [17].

Four (04) figures include Figs. 1, 2, 3, and 4 show bar charts comparing energy consumption of VM allocation algorithms that are normalized with the Tian-MFFDE. None of the algorithms use VM migration techniques, and all of them satisfy the Quality of Service (e.g. the scheduling algorithm provisions maximum of user VM's requested resources). We use total energy consumption as the performance metric for evaluating these VM allocation algorithms.

Simulated results show that, compared with Tian-MFFDE [26] EMinTRE-LDTF can reduce the total energy consumption by average 26.5%. EMinTRE-LDTF also can reduce the total energy consumption in compared with PABFD [4], VBP-Norm-L2 [20] and MinDFT-LDTF, MinDFT-LFT.

6   Conclusions and Future Work

In this paper, we formulated an energy-aware VM allocation problem with multiple resource, fixed interval and non-preemption constraints. We also discussed our key observation in the VM allocation problem, i.e., minimizing total energy consumption is equivalent to minimize the sum of total completion time of all physical machines (PMs). Our proposed algorithm EMinTRE-LDTF can reduce the total energy consumption of the physical servers compared with the state-of-the-art algorithms in simulation results using two (02) parallel workload models [9,17]. Algorithm EMinTRE-LDTF is proved g approximations in general case and $(3 - 2/g)$ in proper interval graphs.

As future work, we are developing EMinTRE-LDTF into a cloud resource management software (e.g. OpenStack Nova Scheduler). Additionally, we are working on IaaS cloud systems with heterogeneous physical servers and job requests consisting of multiple VMs using EPOBF [23]. We are studying the use of Machine Learning techniques to choose the right weights of time and resources (e.g. computing power, physical memory, and network bandwidth).

Acknowledgment. A preliminary version of this work that has been published in the Proceedings of the Future Data and Security Engineering Second International Conference (FDSE 2015). This work was partially supported by the Erasmus Mundus Gate project at the Johannes Kepler University (JKU) Linz, Austria. I am thankful to a.Univ.-Prof. Dr. Josef Küng, JKU Linz for his help.

References

1. Angelelli, E., Filippi, C.: On the complexity of interval scheduling with a resource constraint. Theor. Comput. Sci. **412**(29), 3650–3657 (2011)
2. Barroso, L.A., Clidaras, J., Hölzle, U.: The datacenter as a computer: an introduction to the design of warehouse-scale machines. Synth. Lect. Comput. Archit. **8**(3), 1–154 (2013)
3. Beloglazov, A., Abawajy, J., Buyya, R.: Energy-aware resource allocation heuristics for efficient management of data centers for cloud computing. Future Gener. Comp. Syst. **28**(5), 755–768 (2012)
4. Beloglazov, A., Buyya, R.: Optimal online deterministic algorithms and adaptive heuristics for energy and performance efficient dynamic consolidation of virtual machines in cloud data centers. Concurrency Comput. Pract. Experience **24**(13), 1397–1420 (2012)
5. Beloglazov, A., Buyya, R., Lee, Y.C., Zomaya, A.: A taxonomy and survey of energy-efficient data centers and cloud computing systems. Adv. Comput. **82**, 1–51 (2011)
6. Calheiros, R.N., Ranjan, R., Beloglazov, A., De Rose, C.A.F., Buyya, R.: CloudSim: a toolkit for modeling and simulation of cloud computing environments and evaluation of resource provisioning algorithms. Softw. Pract. Exper. **41**(1), 23–50 (2011)
7. Chen, L., Shen, H.: Consolidating complementary VMs with spatial/temporal-awareness in cloud datacenters. In: IEEE INFOCOM 2014 - IEEE Conference on Computer Communications, pp. 1033–1041. IEEE, April 2014
8. Fan, X., Weber, W.D., Barroso, L.: Power provisioning for a warehouse-sized computer. In: ISCA, pp. 13–23 (2007)
9. Feitelson, D.G.: Packing schemes for gang scheduling. In: Feitelson, D.G., Rudolph, L. (eds.) JSSPP 1996. LNCS, vol. 1162, pp. 89–110. Springer, Heidelberg (1996). doi:10.1007/BFb0022289
10. Feitelson, D.G.: Parallel Workloads Archive. http://www.cs.huji.ac.il/labs/parallel/workload/. Accessed 31 Jan 2014
11. Flammini, M., Monaco, G., Moscardelli, L., Shachnai, H., Shalom, M., Tamir, T., Zaks, S.: Minimizing total busy time in parallel scheduling with application to optical networks. Theor. Comput. Sci. **411**(40–42), 3553–3562 (2010)

12. Garg, S.K., Yeo, C.S., Anandasivam, A., Buyya, R.: Energy-efficient Scheduling of HPC Applications in Cloud Computing Environments. CoRR abs/0909.1146 (2009)

13. Hameed, A., Khoshkbarforoushha, A., Ranjan, R., Jayaraman, P.P., Kolodziej, J., Balaji, P., Zeadally, S., Malluhi, Q.M., Tziritas, N., Vishnu, A., Khan, S.U., Zomaya, A.: A survey and taxonomy on energy efficient resource allocation techniques for cloud computing systems. Computing **98**(7), 751–774 (2014)

14. Knauth, T., Fetzer, C.: Energy-aware scheduling for infrastructure clouds. In: 4th IEEE International Conference on Cloud Computing Technology and Science Proceedings, pp. 58–65. IEEE, December 2012

15. Kovalyov, M.Y., Ng, C., Cheng, T.E.: Fixed interval scheduling: models, applications, computational complexity and algorithms. Eur. J. Oper. Res. **178**(2), 331–342 (2007)

16. Le, K., Bianchini, R., Zhang, J., Jaluria, Y., Meng, J., Nguyen, T.D.: Reducing electricity cost through virtual machine placement in high performance computing clouds. In: SC, p. 22 (2011)

17. Lublin, U., Feitelson, D.G.: The workload on parallel supercomputers: modeling the characteristics of rigid jobs. J. Parallel Distrib. Comput. **63**(11), 1105–1122 (2003)

18. Mastelic, T., Oleksiak, A., Claussen, H., Brandic, I., Pierson, J.M., Vasilakos, A.V.: Cloud computing: survey on energy efficiency. ACM Comput. Surv. **47**(2), 33:1–33:36 (2014)

19. Orgerie, A.C., de Assuncao, M.D., Lefevre, L.: A survey on techniques for improving the energy efficiency of large-scale distributed systems. ACM Comput. Surv. **46**(4), 1–31 (2014)

20. Panigrahy, R., Talwar, K., Uyeda, L., Wieder, U.: Heuristics for Vector Bin Packing. Technical report, Microsoft Research (2011)

21. Quang-Hung, N., Le, D.-K., Thoai, N., Son, N.T.: Heuristics for energy-aware VM allocation in HPC clouds. In: Dang, T.K., Wagner, R., Neuhold, E., Takizawa, M., Küng, J., Thoai, N. (eds.) FDSE 2014. LNCS, vol. 8860, pp. 248–261. Springer, Heidelberg (2014). doi:10.1007/978-3-319-12778-1_19

22. Quang-Hung, N., Thoai, N.: EMinRET: heuristic for energy-aware VM placement with fixed intervals and non-preemption. In: 2015 International Conference on Advanced Computing and Applications (ACOMP), pp. 98–105. IEEE, November 2015

23. Quang-Hung, N., Thoai, N., Son, N.T.: EPOBF: energy efficient allocation of virtual machines in high performance computing cloud. In: Hameurlain, A., Küng, J., Wagner, R., Dang, T.K., Thoai, N. (eds.) TLDKS XVI. LNCS, vol. 8960, pp. 71–86. Springer, Heidelberg (2014). doi:10.1007/978-3-662-45947-8_6

24. Sotomayor, B.: Provisioning Computational Resources Using Virtual Machines and Leases. Ph.D. thesis. University of Chicago (2010)

25. Takouna, I., Dawoud, W., Meinel, C.: Energy efficient scheduling of HPC-jobs on virtualize clusters using host and VM dynamic configuration. Operating Syst. Rev. **46**(2), 19–27 (2012)

26. Tian, W., Yeo, C.S.: Minimizing total busy time in offline parallel scheduling with application to energy efficiency in cloud computing. Concurrency Comput. Pract. Experience **27**(9), 2470–2488 (2013)

27. Viswanathan, H., Lee, E.K., Rodero, I., Pompili, D., Parashar, M., Gamell, M.: Energy-aware application-centric VM allocation for HPC workloads. In: IPDPS Workshops, pp. 890–897 (2011)

Erratum to: User-Centered Design of Geographic Interactive Applications: From High-Level Specification to Code Generation, from Prototypes to Better Specifications

Christophe Marquesuzaà[1]([⊠]), Patrick Etcheverry[1], Sébastien Laborie[1], Thierry Nodenot[1], and The Nhan Luong[2]

[1] Université de Pau et des Pays de l'Adour, Laboratoire d'informatique, EA 3000, 64600 Anglet, France
{christophe.marquesuzaa, patrick.etcheverry, sebastien. laborie, thierry.nodenot}@iutbayonne.univ-pau.fr
[2] Faculty of Computer Science and Engineering, Ho Chi Minh City University of Technology, 268 Ly Thuong Kiet Street, District 10, Ho Chi Minh City, Vietnam
nhan@hcmut.edu.vn

Erratum to:
Chapter "User-Centered Design of Geographic Interactive Applications: From High-Level Specification to Code Generation, from Prototypes to Better Specifications" in:
A. Hameurlain et al. (Eds.): Transactions on Large-Scale Data- and Knowledge-Centered Systems XXXI, LNCS 10140, https://doi.org/10.1007/978-3-662-54173-9_1

The affiliation of the authors Christophe Marquesuzàa, Patrick Etcheverry, Thierry Nodenot, and Sébastien Laborie was incorrectly stated as "LIUPPA – T2iIUT de Bayonne et du Pays Basque, Anglet, France". This has been corrected.

The updated online version of this chapter can be found at
https://doi.org/10.1007/978-3-662-54173-9_1

© Springer-Verlag GmbH Germany 2017
A. Hameurlain et al. (Eds.): TLDKS XXXI, LNCS 10140, p. E1, 2017.
https://doi.org/10.1007/978-3-662-54173-9_7

Author Index

Printed in the United States
By Bookmasters